Making Ontario
Agricultural Colonization and
Landscape Re-creation before the Railway

The colony that became Ontario arose almost spontaneously out of the confusion and uncertainty following the American Revolution, as a quickly chosen refuge for some 10,000 Loyalists who had to leave their former homes. After the War of 1812 settlers began to spread throughout the interlake peninsula that was to become southern Ontario, and by the middle of the nineteenth century expansion had led to a diversifying agriculture and an increasingly open farming landscape that replaced a mature forest ecosystem. The scale of the change from forest to cropland profoundly affected what had been for many decades a rich environment for life forms, from large herbivores down to microscopic creatures. In *Making Ontario* David Wood shows that the most effective agent of change in the first century of Ontario's development was not the locomotive but settlers' attempts to change the forest into agricultural land.

Wood traces the various threads that went into creating a successful farming colony while documenting the sacrifice of the forest ecosystem to the demands of progress – progress that prepared the ground for the railway. Ontario was a going concern before the railway came – the railway simply streamlined the increasing trade with an international market that drew on Ontario for a multitude of farm products and a continuing output from the woods.

Making Ontario provides a detailed look at environmental modification in a time of great changes.

J. DAVID WOOD is professor and chair of the Department of Geography at Atkinson College, York University.

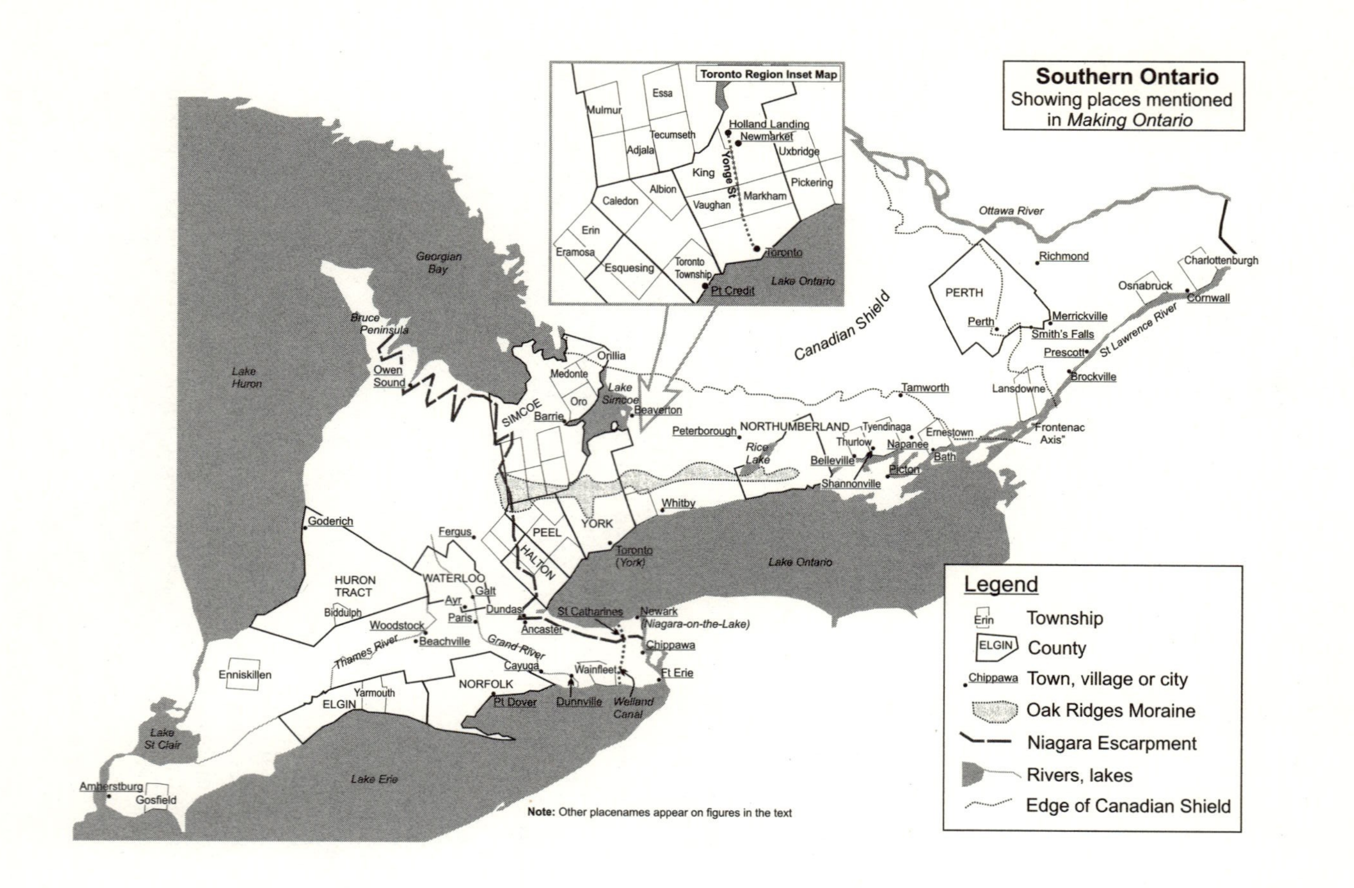
Southern Ontario
Showing places mentioned in *Making Ontario*
Toronto Region Inset Map
Mulmur
Essa
Tecumseth
Adjala
Holland Landing
Newmarket
Uxbridge
King
Yonge St
Pickering
Albion
Caledon
Vaughan
Markham
Erin
Eramosa
Esquesing
Toronto Township
Pt Credit
Toronto
Lake Ontario
Georgian Bay
Bruce Peninsula
Lake Huron
Owen Sound
Orillia
Medonte
Oro
Lake Simcoe
Beaverton
SIMCOE
Barrie
Canadian Shield
Ottawa River
Richmond
PERTH
Perth
Merrickville
Smith's Falls
Prescott
St Lawrence River
Brockville
Osnabruck
Charlottenburgh
Cornwall
Lansdowne
Tamworth
"Frontenac Axis"
Peterborough
NORTHUMBERLAND
Tyendinaga
Ernestown
Thurlow
Napanee
Rice Lake
Belleville
Bath
Picton
Shannonville
Whitby
YORK
PEEL
HALTON
Toronto (York)
Lake Ontario
Goderich
Fergus
HURON TRACT
WATERLOO
Galt
Ayr
Biddulph
Dundas
Paris
Ancaster
St Catharines
Newark (Niagara-on-the-Lake)
Woodstock
Beachville
Grand River
Chippawa
Thames River
Cayuga
Wainfleet
Ft Erie
Enniskillen
NORFOLK
Yarmouth
ELGIN
Pt Dover
Dunnville
Welland Canal
Lake St Clair
Amherstburg
Gosfield
Lake Erie
Note: Other placenames appear on figures in the text
Legend
Erin Township
ELGIN County
Chippawa Town, village or city
Oak Ridges Moraine
Niagara Escarpment
Rivers, lakes
Edge of Canadian Shield

Making Ontario

Agricultural Colonization and Landscape Re-creation before the Railway

J. DAVID WOOD

McGill-Queen's University Press
Montreal & Kingston • London • Ithaca

ISBN 0-7735-1892-4 (cloth)
ISBN 0-7735-2048-1 (paper)

Legal deposit first quarter 2000
Bibliothèque nationale du Québec

Printed in Canada on acid-free paper

This book has been published with the help of a grant from the Humanities and Social Sciences Federation of Canada, using funds provided by the Social Sciences and Humanities Research Council of Canada.

McGill-Queen's University Press acknowledges the financial support of the Government of Canada through the Book Publishing Industry Development Program (BPIDP) for its activities. We also acknowledge the support of the Canada Council for the Arts for our publishing program.

Canadian Cataloguing in Publication Data

Wood, J. David (John David)
Making Ontario: agricultural colonization and landscape re-creation before the railway
Includes bibliographical references and index.
isbn 0-7735-1892-4
1. Land settlement – Ontario – History – 19th century. 2. Agriculture – Ontario – History – 19th century. 3. Landscape changes – Ontario – History – 19th century. 4. Land use – Ontario – History – 19th century. 5. Ontario – History – 19th century. I. Title.

HD319.05W66 2000 304.8′713073 C99-900883-8

This book was typeset by Typo Litho Composition Inc. in 10/12 Baskerville.

For Carnie McArthur (Coull) Wood, John Charles Wood, and Gertrude Amelia (Buckle) Kerpan, Gabriel Stephen Kerpan – pioneers all

Contents

Tables and Figures

TABLES

FIGURES

Acknowledgments

The subject matter of this book has been a perennial theme in my research, and it became the focus of two courses that I have taught over the years – The Pioneer Experience (undergraduate) and The Origins and Patterns of Canadian Settlement (graduate). During this time I have luxuriated in what was an internationally acknowledged hotbed of historical geography in southern Ontario: for most of the past thirty years there have been widely recognized historical geographers active in the universities of Western Ontario, Toronto, Waterloo, Guelph, Wilfrid Laurier, McMaster, Queen's, Carleton, and Trent. At the high point there were seven historical geographers at York University. Although the population of historical geographers is presently diminishing, I have been sustained and inspired over the years by this matrix of kindred spirits.

I am particularly grateful for the example and friendship of John Warkentin, who read and commented on an early version of the book; for the amenable company in my own department in Atkinson College, at York University; and for the intellectual "mateship" of Brian Osborne of Queen's University, Kingston – a toiler in the same vineyard. I am indebted to Jacob Spelt at the University of Toronto not only for a scholarly example but also for allowing me, at a critical point in my graduate study, to venture into the almost untrodden territory of historical geography in Canada; and I was encouraged to go further by Wreford Watson at Edinburgh University, indeed to make a career as a historical geographer. The pioneering work *Canada Before Confederation*, by Cole Harris and John Warkentin, has been a long-standing buttress for my undergraduate course. Tom McIlwraith, in addition to telling me lots about Ontario in *Looking for Old Ontario*, steered me to some useful illustrations. When I have run into "heavy weather" with statistics, I have been bailed out by my expert friend, Michael Cowles, Atkinson College. I have tried out the book at different stages on a few friends, notably Eric Winter, York University, and Bill Tribble, Jim MacPherson, and Earl Winter in Barrie, and have benefitted from their responses. I have had valuable research assistance from graduate students Richard Anderson, Robert Wood, Donna Smith, Greg Finnegan, and Robert Davidson.

It might be seen as curious that, although I have emphasized the debt to my Ontario intellectual environment, the first practical step of actually writing the book was taken in Australia. Thanks to an ideal research environment and technical assistance in the School of History, Philosophy, and Politics at Macquarie University, in Sydney, where I was the 1992 Visiting Fellow in Canadian Studies, I was able to begin word processing the first draft. Both the research activity in Australia and the background for this book have been enriched through grants from the Social Sciences and Humanities Research Council of Canada, from York University, and from the Atkinson College Committee on Research.

With the writing of this book, years of scouring archives and libraries in Ontario and beyond began to materialize as more than just a scholastic enthusiasm. In my searches I have been assisted by many unsung heroes. The library of York University has been my main source of published material, and the conscientious staff, especially in Interlibrary Loans, Government Documents, and the Map Library, deserve accolades. The staff of the Map Library in the Robarts Library, University of Toronto, made available a convenient space and their copy of the De Rottenburg map. Any study of the history of Ontario finds the Baldwin Room, in the Toronto Reference Library, an essential and pleasant place to search; and from time to time I have profitably resorted to my nearby public library in Barrie, Ontario. My most fundamental material has come from archives. I am grateful for the rewarding visits I have been able to make to various collections at the Archives of Ontario. The staff of the Simcoe County Archives (Midhurst, Ontario) has also been helpful. I have benefitted on many occasions from a research visit to the impressive and efficient National Archives, in Ottawa.

Many important points in the text rely heavily on the cartographic renderings of Janet Allin, who drew most of the figures. I am indebted to her cartographic skills and also to those of Fiona Cowles. Dr Richard Anderson exercised his great expertise in computer cartography on the frontispiece map and the finale displayed in figure 8.1.

A section of chapter 3 is a revised version of part of my essay, "Population Change on an Agricultural Frontier: Upper Canada, 1796 to 1841," in *Patterns of the Past: Interpreting Ontario's History*, edited by Roger Hall, William Westfall, and Laura Sefton MacDowell. I thank the Ontario Historical Society for permission to make further use of this material. Also, part of chapter 4 is revised and augmented from my "Between Frontier and Railway: Vignettes of the Social Landscape of Ontario in the 1840s," in *Canada: Geographical Interpretations: Essays in Honour of John Warkentin*, edited by James R. Gibson, Geographical Monographs, no. 22 (Atkinson College, York University, 1993). I am pleased to have received cooperation and permission to include illustrations that are in the collections of the Archives of Ontario, the Toronto Reference Library, and the Royal Ontario Museum.

My family has been embroiled, often unwittingly, in the long process of researching and writing this book. I owe them a tribute and also a recognition of practical contributions toward the completion of the project. I thank Stepan Wood for being my guide on laws and my word-processing wizard and Mary Elizabeth Wood for intelligent research and for assembling the bibliography and the illustrations. Through the long gestation of *Making Ontario*, my wife, Mary, has kept me wonderfully fed and watered, and culturally nurtured. In addition, she has gently applied an editorial scalpel to my lexical infelicities.

I received valuable and thought-provoking comments from the readers for McGill-Queen's University Press and the Social Sciences Federation (Aid to Scholarly Publications Programme). I want to give full credit to the shepherding skills of Joan Harcourt at McGill-Queen's University Press. She stuck with my project through the ups and downs. Rebecca Green undertook the fundamental editorial shaping of the manuscript; she understood remarkably quickly what I wanted to say and, in many places, she helped me to say it more effectively. Joan McGilvray took the book in hand and capably guided it through the technical stages to its final form. In the end, of course, any warts or deficiencies that appear in the work are my responsibility.

Preface

Mid-nineteenth century Ontario was a time of profound and rapid change. We know, for example, that the population grew from 450,000 in 1841 to 952,000 in 1851; we know that temperance societies were proliferating during the 1840s; and it is documented that the quantities of most products shipped into the United States from primarily Ontario sources between 1840 and 1849 increased massively, quite a number of them by more than a thousandfold. What is not so clear is where the striking changes were occurring. How did the appearance of settlement differ from place to place? What were the channels of the land-seeking movement? How did the society function on the land? These questions point to what might be called the *historical where.* This is the orientation of historical geography, and of this inquiry.

Making Ontario was one of the great dramas of the occupation of the New World, and as with all dramas it had a setting, in this case a natural environment free of substantial human modifications. The interaction of settling and environment raises further vital questions: what was the full gamut of geographical change wrought from the 1780s to 1853; and under what ethic did the settlers interact with nature? When Robert Leslie Jones in his classic 1946 work, *The History of Agriculture in Ontario,* called the American-born settlers of early Ontario "land-butchers," he was thinking of their carelessness as farmers, their untidy fields, their ignorance of manure as field dressing. Today we might apply the term differently – these settlers, even the tidy, progressive farmer, were *ecological revolutionaries* who changed the face of the earth.

The interlake peninsula that was to become Ontario[1] was not ecologically homogeneous in the 1780s. There were significant differences in the forest: the littorals of Lake Erie and Lake Ontario were largely covered by hardwood forests, with the interspersion of a variety of more southerly species. In many sandy and gravelly areas, such as north of Long Point, on the crest of the Oak Ridges Moraine, and along the southeast shores of Lake Huron, there were heavy stands of pine. Evergreen species were generally more common toward the north, culminating in large expanses on the sand hills and sand plains to the east and south of Georgian Bay respectively. There were local variations in the forest ranging from estate-like oak plains occurring around the Grand River Valley and through the southwest; blowdowns (some reclaimed by scrub) especially in sandy areas; and scattered trees and thickets on the wetlands that were common across the province. Thus, the natural environment encountered by the Euroamerican settlers was full of variety, a variety well known but little modified by the relatively few aboriginal occupants. Generally, the newcomers perceived the forest to be a monolith against which it was necessary to struggle.

The new landscape being carved out by settlers was similarly full of variety, ranging at mid-nineteenth century from a few townships in which settlement had removed over half of the trees, and a like number of townships that were one-third to one-half cleared, to the majority of townships still typified by fairly continuous forest. A start had been made on removing the forest about the time the first agricultural settlers arrived in the 1780s and woodsmen began taking out choice timber for export through Quebec City.[2]

The romance of the railway, and the copious attention it has received over nearly two centuries, has led to the superficial notion that the train was the major transforming force in the New World in the nineteenth century. One has to recognize the powerful influence of the railway in the reorganization of the economy and the emergence of urban dominance by the beginning of the twentieth century, but in tracing back the roots of Ontario and measuring geographic change, it is clear that the primary instrument of transformation through nearly the first century of making Ontario was not the locomotive, but the axe. The domain of southern Ontario was transformed from one ecologic category to another – from woodland to farmland – in less than a hundred years by an army of axe-wielding settlers and woodsmen.

A geographical analysis naturally raises questions about the locations of developments, variations from place to place, cultural and political expressions in the landscape, and the characteristics and involvement of the natural environment. What was the process and pace of land clearing and what was the impact on the natural environment? What was the role of agriculture in this example of New World settlement, and to what extent was it expressed in political influence? How did the pioneer population fare, and how did people organize their society and livelihood? To what extent was Ontario a typical frontier society, and to what extent did it illustrate (or belie) the New World experience? If, indeed, this book uncovers the roots of what is modern Ontario, at what point do we begin to see the crystallizing of a recognizable structure? The overarching context of all these questions and of the documented metamorphosis of the first few decades under farm settlement is the massive transformation of a part of the eastern woodland of North America.

Various fundamental trends that influenced the conditions and character of the province during the bustling 1840s and early 1850s are traced from earlier rudiments and give substance to the themes in the book. These include the ecological onslaught caused by agricultural settling and timber extraction; the modulations and movements of the pioneer population over a half-century; the slow crystallization of social facilities and organization, both conceptually and concretely; the gradual emergence of a broadening export agriculture from the domination of timber and wheat; and the evolution of transportation and communications media in concert with an embryonic urban network.

Over the past century historical geography has developed a number of effective techniques of analysis, the classic model of which is a combination of chronology and cross-section. A cross-section usually elaborates a theme or an area at a point in time, whereas chronology follows a theme over time.[3] This book honours the traditions of historical geography in its structure, proceeding chronologically, but regularly highlighting a theme at a significant time, especially through the maps and tables.

Historical geography, along with human geography (its academic home) is presently concerned with the political and philosophical quicksand of the language we use to articulate our questions and understandings. The discipline is in a period of soul-searching and reassessment of the ways in which its research has been done and its knowledge has been communicated. Rethinking the past is going on apace, and identifying and correcting omissions (old and new) is the order of the day. A realization of what we have done to the natural world has gripped the growing body of reassessments of colonial settlement. A recent volume expresses concern with "ecologies of invasion" and suggests an "ecological apocalypse."[4] There is probably little to be gained by being penitential, but understanding the scope of the changes we have caused is a good basis for judging prevailing values and inspiring plans for amelioration.

This book is not "the last word." Indeed, there is an increasing body of scholarship, much of it to be found in the footnotes herein, that enlarges on themes related to this book. What is attempted here is a contribution based on the perspectives and research strengths of historical geography.

INQUIRING ABOUT ONTARIO BEFORE THE RAILWAY

This book is about a process of gain and loss. What was gained was the material progress of tens of thousands of settlers, especially those that took on a piece of wild land and set about taming it for agriculture. What was lost was the awesome complex of eastern woodland plants and animals that stood in the way of agriculture. Visualizing the process as ongoing from the 1780s, when the first American Loyalists settled on the north side of the Great Lakes system, and gathering momentum noticeably after the War of 1812, this book traces the components of change up to 1853, as the first trains began to move across Ontario's landscape.

Chapter 1 sets the scene for this process by discussing "progress" as an ideology that encouraged activities of settlement, and by exploring the rhetoric that settlers used in confronting their environment. The massive landscape transformation, sketched out in chapter 2, provides a context for the various social and economic developments that are dealt with in the succeeding chapters.

Chapter 3 begins placing the "players" in the landscape by reconstructing the population characteristics of the pioneer society, continuing through the great changes and rising numbers of immigrants after the War of 1812 to the time of large rural families in the productive farmscapes of many of the older settlements around mid-century. As population increased, the area occupied by farm settlers expanded, and population density rose in many of the long-settled areas. Various demographic features are discussed, and the population bases for regional differences are explored.

Chapter 4 shows how this population gradually created the fabric of a modernizing society in a setting that rapidly lost the characteristics of wilderness. The spread of various social organizations across the province, and evidence of social dysfunctions (i.e., the normal gamut of a society) are illustrated.

Chapter 5 lays out the economic framework on which the society crystallized, from the simple exploitation of staples, under the fading mercantilist system, to the metamorphosing agricultural economy and landscape on the eve of the inauguration of the railway. In chapters 5, 6, and 7, the beginnings of industries other than those based purely on agriculture are traced. Many establishments engaged in designing and building larger pieces of farm equipment, such as foundries or carriage makers, shifted into building steam power equipment or other complex machines.

Movement around the province continued to be difficult until at least the 1840s. Chapter 6 clarifies aspects of the circulation of goods and people on the ground and by water. Transportation facilities were the heaviest capitalized developments both before and after the middle of the century. Until the 1840s information and ideas circulated primarily by word of mouth (later called the "bush telegraph") and through the post office, bolstered by handwriting and a few printed sources, such as newspapers. With the telegraph, in the 1840s, came the first spark of what was to be the electronic revolution.

Chapter 7 depicts the roles of urban nuclei in the maturing settlement of Ontario and the evolving *mise en scène* from a multitude of hamlets and a few larger places to an embryonic urban network. Specific examples of the functional complexes of a number of urban places are also laid out.

Chapter 8 reflects on the gains and losses of what was a major metamorphosis – arguably, an ecological revolution – in Ontario over the first two or three generations of agricultural settlement. The changes that had been wrought in just over half a century made Ontario economically promising enough to attract international capital for major projects, especially for the building of railways. A new era was about to begin.

Lanark County stone house built in 1830 by Scottish immigrant John Baird (Archives of Ontario [AO], Acc.2449, s5698)

Perth, the Capital of the District of Dalhousie, from the N.East bank of the River Tay: sketched 20th Aug. 1828 (AO, Thomas Burrowes fonds, C1-0-0-0-22)

Richmond on the River Jacques, or Goodwood, a tributary of the Rideau River: Sept. 1830 (AO, Thomas Burrowes fonds, C1-0-0-0- 117)

Brewer's Lower Mill; view down the Cataraqui Creek, & Clearing made for the Canal. Sketch taken in 1829, Excavation for the Lock just commenced (AO, Thomas Burrowes fonds, C1-0-0-0-67)

Bath on the Bay of Quinte, and Upper Gap to Lake Ontario. 1830 (AO, Thomas Burrowes fonds, C1-0-0-0-95)

Saw Mill, &c. on the Salmon River, at Shannonville [ca 1830] (AO, Thomas Burrowes fonds, C1-0-0-0-107)

Head of the Bay of Quinte, from the Carrying Place [ca 1830] (AO, Thomas Burrowes fonds, C1-0-0-0-111)

Fergus, Upper Canada. Hand-coloured lithograph after Miss J.D. Fordyce, ca 1835 (Royal Ontario Museum, 75 CAN 371, 969.129)

Corduroy road over a swamp in Orillia Township, Sept. 1844, by Titus H. Ware (Toronto Reference Library, Acc. M1-17, T 14377)

Scene on Lake Scugog, from *The Anglo-American Magazine*, Jan. 1854 (Royal Ontario Museum, 88 CAN 15, 981.55.15)

Making Ontario

1 "Progress" and the Confrontation with Nature

> This is an age of progress ... The tide of population and production which is destined to flow over and to fill the several channels of communication as they are successively opened up, is rising higher and higher on every side ... It would seem to be ... the part of true wisdom to look with much confidence to the future, and to take due account of advantages which, although they be prospective, are by no means problematical.
>
> Lord Elgin, 1851[1]

This is the rhetoric of progress in a "new era" for Ontario (Canada West), as proclaimed by the Queen's representative at the sod-turning of the Ontario, Simcoe, and Huron Railway. Only twelve years earlier a different governor general had spoken of two nations warring in the bosom of a single state – and, with specific reference to Upper Canada, of "a monopoly of power and profit."[2] That representative, the controversial Lord Durham, had been dispatched to Canada to investigate the Rebellions of 1837, which, in Upper Canada at least, had been triggered by perceived barriers put in the way of ambitious, frustrated farmers by an entrenched, reactionary establishment.[3] This was not a new situation. For over half a century, Ontario had struggled as an agricultural settler colony with an elite, colonial oligarchy that governed with its own interests in mind rather than those of the farmers.

There had been serious setbacks for the merchants and the administrative life of the colony over the years. Upper Canada had moved to the outer fringe of concerns of the home government during most of the years of Napoleon's campaigns, and growth was very sluggish at the beginning of the nineteenth century. (Whereas nearly one hundred townships were surveyed between 1783 and 1799, only forty-one were surveyed between 1800 and 1815.) The three years of the War of 1812 were times of turmoil when many activities were "on hold" or greatly restricted. The 1820s opened with a rapid increase of immigrants from overseas, which led to class tensions and ideas of social change. Although the period around 1832 was marked by heavy immigration, with a population increase averaging 7 per cent a year, and major development proposals, it was also marred by invasions of cholera (1832 and 1834) and, at the end of the decade, appearances of feared wheat pests.[4] During the later 1830s the economy slipped into recession (along with that of the United States), and the years 1837–38 were tumultuous times of armed uprisings in many rural areas. The uprisings directed imperial attention to the colony once again, and the 1840s brought attempts to ameliorate conditions and encourage a larger degree of independence.

Considerable apprehension was raised in the colonies by the repeal in 1846 of the Corn Laws that allowed Canadian wheat to enter the British market under a preferential tariff. But Upper Canada was finding other markets for its products, and the 1840s proved to be a period of great activity and innovation on many fronts – a kind of fusion of many credible ideas with increasingly available capital, much of it foreign. Such capitalization had not been possible in previous decades of locally focused and poorly connected interests. Industries more sophisticated than saw and grist mills began to appear, and toward the middle of the century some promising railway proposals were planned, laid out, and financed.

A COLONY FOR FARM SETTLERS

If early Ontario was an agricultural frontier – and there are no grounds for denying it – it did not fit the popular conception of continuous and universal frontier progress. There had been many occasions over the sixty years following the Loyalist settlement when different interest groups were at odds. A chasm often existed between a farming population that – rightly, it would seem – expected the administration of their primarily agricultural colony to aid their struggle to clear and farm and a governing oligarchy whose main concerns were not with farming. Around 1820, the irascible Robert Gourlay put words to the widespread frustration of his fellow settlers: "the actual settlers, up to 1817, remained in little communities, cut off from each other ... and left imprisoned in the woods. They cannot dispose of their farms: they cannot afford to abandon them; and they pine on." This added to what Gourlay called "the present very unprofitable and comfortless condition."[5] That condition was slowly modified during the 1820s and into the 1830s, though not always for the better. By the 1840s Ontario was spoiling for change, and it came in many forms.

As Gourlay and others came to realize, the major political power during the early decades of the colony did not reside with farmers, despite the overwhelming importance of agriculture. Until after the War of 1812 the main governing body was the appointed Legislative Council, led by its executive core group. The appointees to the council were men considered by the lieutenant-governor to be the chief citizens of the colony, and the pillars of the colony as the colonial administration envisaged it. Many of them had desirable posts in government, or in institutions related to government, and they had a vested interest in maintaining the status quo. They were town dwellers, not farmers, although they were likely to own large acreages (witness the Family Compact and various "patrons" across the province).[6]

The elected House of Assembly gained more influence in governance from the 1820s. As J.K. Johnson illustrates, assembly members were much closer to the people than council representatives, but active farmers seldom formed a majority. Of the 283 members who sat in the assembly during the fifty years ending in 1840, just over half were farmers at some time in their careers. It is not surprising that the most prominent farmer members of the House of Assembly were Reformers.[7]

The growth of population, changing representation, and concomitant legislative activity in the growing assembly led to a gradual diversification of political power and to the evolution of a working administrative structure for the society. At the beginning of the 1820s, the main socio-political entity in the day-to-day life of a settler was the township (as it had been since the 1790s). At the next level was the appointed periodic Court of General Quarter Sessions that dispensed the law and general administration, and, beyond that, a superstructure of courts and governance, most of which had been transported from England with little modification. This framework was later to give weight to the clamour for "the rights of Englishmen," and by the 1840s, for responsible government. Township government, with its annual election of local officials, had thoroughly evolved in North America, and was often seen by colonial officials as tainted with democracy, and thus a threat to the status quo. According to C.F.J. Whebell, the townships maintained or even added to their influence, within their own boundaries, until the late 1830s. Then, in 1841, by the District Councils Act, many of their responsibilities and many of the petitions formerly sent to the assembly were redirected to the districts.[8]

Through a combination of imported structures and indigenous needs, a social system emerged. From about 1830, social institutions of many kinds proliferated. There was an increase during the 1830s, and more so during the 1840s, of agricultural societies, temperance societies, churches, mechanics' institutes, and fraternal organizations. As they evolved, these groups not only infused the society with personal interconnections but also became involved, through the buildings they erected, in creating parts of the landscape. For the common settler, long-distance communication with loved ones was conducted through the postal system; hence, a post office came to be perhaps the most yearned-for feature of a new settlement. Because the need was carefully assessed, the spawning of post offices and the delivery of mail provided concrete evidence of a crystallizing society – of progress.

In addition to being a place of refuge for the Loyalists, Upper Canada was to be a settler colony based on agriculture. Timber exploitation, a remnant of the empire of plantations and mercantilism, was a separate activity from farming, except insofar as farmers were glad to sell trees they removed from their farms in the clearing process. From time to time certain areas destined for farming temporarily became major producers of timber for external markets, as did the area north of Lake Erie and Lake St Clair in the 1840s. It has been calculated that there was a saw mill for every 700 people in New York State in 1855.[9] The average for the province was one saw mill for 608 people,[10] but there were many parts of Ontario in 1851 where the ratio was much higher: Norfolk County had one saw mill for every 217 people, Elgin County one for every 322, York County one for every 302, and Northumberland County one for every 380. Nevertheless, from an early date Upper Canada was destined as farming territory, an expectation similar to that for the United States, where Benjamin Franklin claimed that agriculture was the "great Business of the Continent."[11] An economy based on farming was the foundation of the progress that Lord Elgin lauded in 1851. It was also the primary basis of the potential that was looked for in deciding on a post office location.

In granting or refusing an application for a post office in 1841, Deputy Postmaster General T.A. Stayner assessed the tributary population and the promise of the settlement. He describes the application process as follows:

> The inhabitants of the place where the Post Office is needed, address me by Memorial or Letter, stating their want, and recommending a person as Post Master; the merits of the application are then enquired into by the Surveyor, who visits the place indicated for the Office, ascertains the fitness of the individual nominated as Post Master, and reports the result of his investigation to me. By this report I am in most cases governed; but I beg to observe that the question of creating a new Post Office, or changing a Post route, or increasing the travel of the Mail upon a line of post, frequently involves a necessity for much research and correspondence, and that there is no branch of my duty which calls more forcibly for the exercise of the best judgment I can command than this.[12]

As the population grew and fanned out into newly surveyed townships, the cleared areas gradually coalesced and were pinned together by rough roads and the burgeoning of post offices and other community structures. As in all parts of the world, the increase in the number of people was the chief cause of the rising pressure on the natural world. Along with the increasing population and a materializing social system went the building of an economic system. Early in the nineteenth century, Ontario began to fulfill an imperial role as a supplier of large timber and some grain. Forest products of various kinds continued to be the most important component of Ontario's export trade until well into the second half of the century. In 1852 manufacturing was still an underdeveloped activity for a population that had reached one million. Agriculture was by far the dominant economic sector, although in the 1840s substantial diversification into what James M. Gilmour identifies as consumer and producer goods, and a broadening service sector, had begun to occur.[13] It was almost entirely as a result of the economic success of agriculture (and the availability of imported capital) that the first Ontario railways were built. By the middle of the century, Ontario was a major agricultural producer in international terms, comparing favourably with the most productive part of the United States at the time.[14]

The increasing rural population in Ontario soon demanded certain aids to agricultural livelihood and social interaction. Many proto-urban nuclei that offered a service to the pioneering farm population appeared – perhaps a mill, or a blacksmith's forge, or even a general store. As time passed and conditions changed, many of these nuclei proved to be ephemeral. For those that survived there was a common pattern: in the beginning of an agricultural settlement, the proto-urban nucleus had simple, repetitive relations, primarily with its immediate hinterland; as population increased and development proceeded, different products and skills were called for, and relevant, complementary businesses began to cluster in the vicinity. Improving roads linked market centres and sources of supplies to their clientele. On the eve of the introduction of railways (in Ontario, about 1850), there was a roughly defined urban hierarchy having, at most, three tiers: major centres, regional centres, and small centres. The half dozen major centres, were usually ports and transshipment points, or concentrations of government functions. Tributary to particular major centres, the regional centres were relatively few in number. A range of services, perhaps including government, was available, and goods likely changed hands here before going further. Small centres, tributary to the regional centres offered a truncated selection of services and had no business connections with other small centres.[15] The small centres provided what was needed by a straggling, pioneering population engaged in common efforts to turn woodland into farmland: farming tools and repairs, building materials, and mills for transforming produce.

The introduction of the railway brought change to all urban places: within thirty years it would traumatize the third tier (small) centres, undercutting and isolating many places not on the railway line. (In fact the practice of moving buildings and even whole villages to be near the railway line began early in the railway era.) The railway would challenge all the regional centres and reinforce some, and it would magnify the roles of almost all the major centres. It should be remembered that in pre-industrial Ontario, unlike the present, urban places were almost entirely offshoots of farming activity (though resorted to only irregularly by most rural dwellers). The pre-eminence of urban influences today might tempt one to think of the proto-urban nuclei of early Ontario erroneously as small engines of progress just waiting to spring into action. In fact, urban places served the farming way of life in general: what mattered most was what was happening in the country rather than in the village. Once the farm landscape was well cleared and the agricultural economy established, many of the small nuclei became less necessary and, by the end of the nineteenth century, began to wither away.

THE LANGUAGE OF LANDSCAPE CHANGE

Agriculture has been the most universal human imprint on earth, and, unlike the largely symbiotic food gathering done by primitive groups, it has resulted in a profound and thorough transformation of the land. Agriculture had an unequivocal impact on Ontario, where it replaced a neatly integrated woodland ecology with a simpler and largely exotic – in the sense that most of the crops were not native and the biological diversity was lower on cropland – land use regimen. This was not a benign exchange: terms of warfare were commonly employed. Many visitors echoed William Smith's observation, at mid-century, on the Oak Ridges Moraine: "the universal Canadian practice has been followed in clearing the land, that of sweeping away every thing capable of bearing a green leaf ... The new settler ... looks upon trees as enemies."[16]

Underlying the various signs of progress in the settler colony – in demography, social structure, economy, communications, and urban germination – was the rhetoric of struggle with a vigorous, long-standing natural environment. In many ways, this rhetoric was the product of a technologically aggressive, exploitative culture, reaping resources in many parts of the world. It gave voice to the prevailing view that the changes wrought in Ontario between the 1780s and 1853 were the products of an epic victory of human ingenuity and effort over a challenging wilderness.

The praises of "Glorious Nature" and "the monarchs of the forest" by observers from the Old World and a few local aesthetes could not be comprehended on the frontier. Richard Rorke, an Irish Quaker who emigrated to Ontario as a young man, fills his book of reminiscences with tales of tracking bears, trapping wolves, and killing many forms of wildlife, suggesting that, in those days, even people who preached peace did not extend it beyond humankind.[17] The general ethos in rural North America was anti-nature. As one memorialist said, "the mainspring in my father's life was a desire to subdue wild lands. His triumphs were those of man over nature."[18] Even for most overseas settlers from a tame home country, the marvel and romance of the wild was soon pushed aside in favour of the farmer's need to take up arms against the encroaching vegetation and predatory animals. Settlers could not afford to be tourists: for them, beauty lay in gaining control over the surrounding natural environment. In Denis Cosgrove's trenchant phrasing, "for the insider, there is no clear separation of self from scene."[19] So immigrants and native-born alike set to engineering an ecological revolution, in which the most essential weapon was the axe.

A much less prominent, dissenting view was moderated by familiarity with the Romantic literary depiction of nature popular in the old country. These juxtaposed views suggest two camps: one composed of insiders (or settlers), the other of outsiders (or visitors). The insiders were broadly engaged in survival, in carving out a livelihood, and they knew first-hand about the threats to personal or group well-being from wild animals and natural hazards. The outsiders were, by and large, passing through – touring – and were therefore able to achieve separation from the scene, as Cosgrove puts it. The insiders grew up with and assimilated a lexicon that reinforced their self-confidence and courage. As the early environmentalist, John Muir, recorded in Simcoe County in the 1860s, "so many acres chopped is their motto, so they grub away amid the smoke of magnificent forest trees, black as demons and material as the soil they move upon."[20] By using an aggressive, heroic rhetoric that demonized the dark forces of nature, they set out to overcome the almost imponderable environmental barriers to making a farm: they gained backbone for the settlement process.

It became widely accepted by residents of early Ontario that it was within one's rights to deal roughly with the natural environment. There was little sensitivity to the often unique species in the woodland cornucopia or to the originally abundant wildlife. For the established pioneer, shooting wild animals, even those animals not competing with or threatening domestic livestock, evolved from the early experiences of confrontation to a form of sport. Richard Rorke's table of contents glorifies the hunter and the chase: "My First Bear Chase," "An Attack At Midnight," "Stealing a March on the Enemy" (all versus a bear), "The Wolf, The Wolf," "A Stern Chase Is a Long Chase" (involving a wolf dragging a trap), "To Drive the Deer With Hound and Horn," and "On the Track" ("the usual strong inclination came on me to kill a deer").[21] In the human/nature relationship, the human is the exploiter and nature is the object to be exploited, with no second thought for rights or mores.[22]

The rhetoric of confronting and defeating wildness was a male prerogative. In fact, women who wrote about such things apparently eschewed the aggressive stance as not appropriate for them. Being fearful and awestruck were more typical responses in women's writing, and Patricia Jasen argues that this was a result of the kind of literary expression the public expected from women.[23]

Combatting nature was an integral part of progress in the nineteenth-century mind, and anything that could be labelled "progress" automatically won out over preserving a corner of the natural world.[24] The image of progress as armed opposition to nature became deeply embedded in the nineteenth-century psyche, and only slowly softened through the generations. (Indeed, this notion still exerts considerable influence, despite different conditions and much greater knowledge, in the current jobs versus environment debate.) Progress was a kind of *superorganic* term, in that there was more to it than indicated by a dictionary definition, like certain reverberant words in currency today such as "postmodern."[25] The ideology of progress was the justification for the policies of liberal individualism that imbued the society; together, they attacked the natural environment through language and action.

In contrast to the insiders, most of whom regularly faced the challenges of nature, the outsiders expressed themselves through a different rhetorical tradition, with language more appropriate to the genteel drawing-room. Most outsiders were short-term residents, or tourists, but there were some long-term residents who achieved a measure of separation from the scene. Usually these were upper-middle-class people with an external source of income, such as a pension. These touring outsiders, whether male or female – and a large proportion were women – seem to have had no quarrel with the preeminence of progress and the necessity of clearing land for agriculture. In fact, the males in this camp generally celebrated the achievements of the landscape transformation, and some, including John Howison, ridiculed farmers who were less assiduous in their clearing.[26] Most outsiders were looking for the unusual, and could afford to voice admiration for the wonders of nature. Since they had protection from both the competition of wild predators and the toil of making cropland, their language could deal more with the aesthetic and the marvelous aspects of nature. La Rochefoucauld-Liancourt, in 1795, says that "the prospect of the Grand Cataract of Niagara [was] one of the principal objects of our journey"; and John Goldie, the botanist, while claiming disappointment, spent seven of a total of sixty pages of his diary in 1819 describing Niagara Falls and its surroundings.[27] As members of the middle and upper classes, these observers were affluent enough to travel, and had been trained to look for the beautiful and the unusual, and to describe in tasteful ways what they saw. For the visitor, it was the natural features or rural scenes that caught the attention: things urban, if mentioned at all, were usually compared unfavourably with towns elsewhere.

There were also other voices, many from those immigrants who settled in town and found a way of making a living other than farming. For example, William Smith's text periodically breaks away from the dispassionate account of population numbers, production, and notable structures to mourn the loss of noble trees or comment on the mindless slaughter of wildlife.[28] Perhaps there was also a general difference in attitude toward nature between town and country dwellers, the former being somewhat insulated against the hazards of the forest.

Although some might claim that the recounting of human/nature relations is innocent of ideology, there is no denying that the natural environment of Ontario was profoundly and *purposefully* changed during the century after 1780. That transformation, the drive for progress, in itself became an ideology – indeed, the prevailing, almost universal land ethic of the province.

2 Changing the Face of the Earth

> Man has disturbed and displaced more and more of the organic world, has become ... the ecologic dominant, and has affected the course of organic evolution ... He has worked surficial changes as to terrain, soil, and the waters on the land.
>
> Carl O. Sauer, 1956[1]

It takes little reflection to realize that the activities taking place on the farms of nineteenth-century Ontario – cutting and rooting out trees, burning debris from clearing, draining wetlands, driving off or killing wildlife, erecting structures – would have had a profound environmental impact. The timber exploitation had perhaps an even greater effect on the waterways – at least before the main canal building – through the scouring and gouging of river banks by rolled or dragged logs; damming and diverting of streams to expedite the logrush; and sedimenting of lakes and streams with eroded soil, "deadheads," and debris from felled trees. In fact, all facets of the settlement of early Ontario by Euroamericans were gradually transforming a forested landscape into an open landscape that was regimented by the straight lines and right angles of the geometry imposed by humans.[2]

THE MECHANICS OF TRANSFORMATION

The pioneer farmers were engaged in organizing their properties to make them easily usable and productive. Making them look like the farms of their home areas was probably of less concern, but, of course, the settlers' ideas of how best to organize a farm came from previous experience, as did their knowledge of husbandry.[3] A vitally important means of modifying the environment was the importation of seeds from overseas. Settlers often brought seeds with them, and relatives or friends who came to join them later were commonly instructed to bring certain desirable propagules. These imported plant materials included useful food crops, on one hand, but also some of the major weed scourges, on the other, such as thistles, spurge, burdock, and wild mustard. Most of the plants identified as weeds today are natives of Europe or Asia, not North America. It is said that the Scotch thistle was introduced by a homesick settler. This process, described as "ecological imperialism" by Alfred Crosby,[4] is illustrated by the letters of Joseph Carrothers. Writing from near Lake Huron to Ireland, he asked for certain plants to be sent. The transfer was not entirely one-sided: Carrothers also sent seeds to Ireland.[5] Another good example is a letter from James Sharp, who wrote in 1821 from Dumfries Township, Ontario, to relatives in Scotland, asking for seeds of ryegrass, the long pod bean, early cabbage, cauliflower, yellow turnip, good oats and barley, and a particular Scottish potato. Sharp complained that Ontario "Oates is the poorest trash you ever Saw."[6] Many of the major advances in food crop breeding were based on imported plant materials.

The gradual obliteration of the forest was the most obvious aspect of the transformation of the landscape. The rhetoric of progress and the proliferation of saw mills, as discussed in the previous chapter, provided, respectively, a philosophical and an operational framework for the land clearing. Although only some townships along the Lake Ontario shore and in the Niagara Peninsula were more than half cleared by the middle of the century, and almost everywhere the scene was a patchwork because of varying rates of clearing, the removal of the trees was proceeding inexorably. Whether or not all Ontario farmers viewed trees as enemies (as some visitors claimed), the cutting of trees went on: the land had to be fairly clear for the planting of food crops; the fuel for heating houses and powering steam engines was wood; and, where smelting was carried on, large amounts of wood were required for making charcoal.[7]

After one full generation of clearing, wood for fuel became the chief demand on the woodlands. It was quite possible that a new settler's first "crop," in addition to wood ashes for potash, might be wood for fuel or charcoal burning. In his magisterial work on the forests of the United States, Michael Williams suggests that the amount of wood used for fuel up to 1810 "far exceeds the total amount of lumber cut."[8] (Reflecting comparable conditions, the demand for both lumber and fuel in Upper Canada increased exponentially in the following generation.) William Cronon also highlights the use of wood for fuel in New England, which compares well climatically with Ontario: "a typical … household probably consumed as much as thirty or forty cords of firewood per year, … [which] meant cutting more than an acre of forest each year. In 1800, the region burned perhaps eighteen times more wood for fuel than it cut for lumber."[9]

Because of the initial abundance of supply, Americans tied almost all their energy needs and technology (apart from water power) to wood fuel. This would have been true for Ontarians as well. It is reasonable to assume that the heating of buildings and fueling of industries in Ontario during the first half of the nineteenth century would have mounted as great an assault on the woodlands as the taking of trees for lumber, including lumber for export.[10] Even without the destruction by land clearing or the sale of trees for lumber or charcoal, the wood available on most 100- or 200-acre farms would have been rapidly diminishing by the middle of the century (see figure 2.1). Williams estimates, that a moderate yield of fuel wood from one acre of forest around 1850 would have been twenty cords, one bush cord being four feet high, four feet wide, and eight feet long. A simple calculation, assuming a low estimate of twenty cords of fuel wood per household per year, shows that, in 1842, Ontario would have required over 1,500,000 cords, or the yield from about 80,000 acres of woodland. By 1851 there were 50 per cent more households; therefore, during the 1840s, the fuel wood demand would have required the equivalent of one and a half to two large townships each year, for a total of about one million acres over the decade.

There were instances in which war on the tree cover would have been an appropriate description indeed. Many early settlers had a cavalier attitude toward the timber on Crown land, and the pilfering of trees was a widespread problem. As the demand for lumber increased, especially for export to the growing cities in the United States, the government had to take action to protect public woodlands. From time to time various tricks were employed by bogus settlers, such as making a small down payment on a farm lot, then removing all the valuable timber, selling it, and disappearing.[11] There were cases of invasion of public woodlands by experienced woodsmen, who would rapidly cut down the timber, sell it, and move on, sometimes with the apparent connivance of established timber merchants. Correspondence to the commissioner of Crown lands gives a snapshot of this perennial concern:

York 23d Jan'y 1829

Sir,

I have the honor to report that in obedience to your instruction of the 23d May 1828 ... I have Inspected the Township of Tyendinaga in the Midland District and found only thirty eight Settlers ... residing upon the lands, seven eighths of which are Squatters ...

There is a considerable quantity of Squared Oak and pine Timber in the Township now ready for drawing [removing], also a quantity of Oak Staves and a great number of Pine fitt [?] for Saw logs many of which are cut ready for

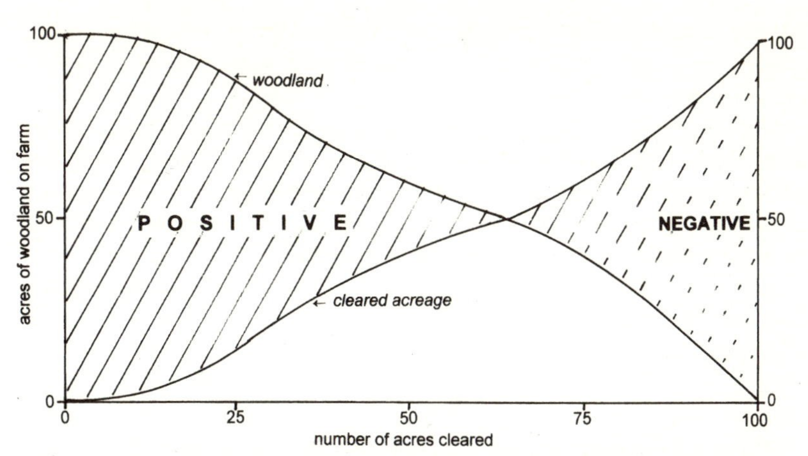

Figure 2.1 The relationship between land clearing and fuel wood supply on a 100-acre (40 ha) farm. The farther the clearing extends from the main buildings, the more inconvenient the wood supply, until it becomes necessary (probably in the second generation) to buy wood from sources off the farm. Gradually removing the tree cover changes from a positive farm-making activity to the exhausting of a needed resource.

drawing; and [I] should recommend that an officer should be sent down without delay to make a Seizure [?] otherwise the Lumber will be on ice[.] this has been drawn off as they where [*sic*] only waiting my absence to commence.

There are many persons that have taken possession of lots by only felling three or four Trees this is in consequence of a report from this place "that any person having taken possession of a lot will not be dispossed [*sic*] by the Government" which practice ought not to be allowed as it will prevent good Settlers from purchasing.

... Sam.[l] S. Wilmot
Dy Surveyor

* * *

Township of Toronto
18th May 1829
Sir,

Agreeable to your instructions of the 11th Ins.t I came to this Township to day – and have seized about Two Thousand five Hundred Staves, culled and marked T.S. (Timothy Street of Streetsville) made by Patrick Tobin – with a number of trees felled & marked T.S. – About 3000 were made by Robert Jaynes and 2000 more by John Falconer – the latter 5000 appear to have been removed during the frost. – I have placed the Staves and timber (on the lot) in charge of George Falconer – and have instructed him, not to allow them to be removed until he hears from you – or myself – The 2000 Staves made by John Falconer the timber for which, he purchased from Robert Jaynes and Patt'k Tobin Neither of them possess real estate – and as I am informed make a business of Trespassing upon the Reserved Lands – John and James Falconer would answer for witnesses together with Timothy Street who purchased the Staves from them.

Have the goodness to acquaint me at your earliest convenience what further steps I am to take in this business – It would have been useless to have looked after such staves as have been removed – in all probability they are at the mouth of the Credit [River] – and perhaps across the lake – However the quantity can be proved against them. I have the Honor to be Sir ...

William J. Kerr

[Sent to] The Hon'ble
Peter Robinson
York[12]

There were different kinds of repercussions from the consuming of the forests and the concomitant destruction of most of the previous ecological relationships. The issue that received the greatest amount of debate in the nineteenth century was the effect of clearing on the climate, although, apart from intimations by some visitors, this was primarily a debate of the last half of the century.[13]

The effect on the climate at large of the clearing away of millions of acres of trees may still be debatable, but it has been demonstrated conclusively that there are microclimatic repercussions. Open fields are vulnerable to the drying heat of the summer sun and the full force of the wind, and thus to the blowing of soil. Woodlands moderate temperature change, both in heat and cold, and maintain higher humidity than open fields; thus, farmland heats up and cools down more quickly than tree-covered terrain. The extent of clearing varied from place to place, just as the personal experiences of settlers varied (see figure 8.1, which summarizes the extent of transformation).

It took longer for observers to become concerned about other ramifications of removing the trees, such as the washing of topsoil off cleared fields. There has been a growing recognition in the past century and a half that removal of the tree cover leads to more rapid runoff (thus to more frequent, higher floods), the excessive deposition of sediment in stream beds, the seasonal drying up of headwaters because of the loss of the "sponge" function of woodland, the lowering of underground aquifers, the quicker spread of crop diseases, and the disappearance of much wildlife.[14] A related result of the erosion of the soils during nearly two centuries is the virtual impossibility of describing the characteristics of the soils the settlers found. The original surveyors' descriptions do not coincide well with modern soil surveys. This suggests that the basic qualities of the soils have been modified since the 1780s.

Contemporary publications offered advice on how to judge the quality of the land for agriculture from the tree species. The ability of the average settler to make use of this advice can easily be exaggerated. For one thing, the settlers from overseas, even if familiar with European trees, were not familiar with North American varieties or species. Ontario woodlands were called "mixed forest" for good reason. Even a knowledgeable settler trying to judge one or two hundred acres of typically variable land based on a few dozen yards next to the survey line would have had mixed success. Except for extremes, such as solid stands of white pine, the judging of heavily wooded land was more successful when based on aspect, roughness, and apparent availability of potable water along with good drainage.

The exploitation of the woods for lumber had another gradual but profoundly destructive result. Saw mills located on streams discarded sawdust and wood debris into the convenient waterway. Over time, this material clogged parts of the stream and coated spawning grounds. The dumping of saw mill waste, along with the dams and weirs built across streams for water power, caused devastation of the fish population, notably the salmon. By the middle of the century, the salmon were faced with greatly reduced opportunities to spawn, and this food source, on which all classes had feasted in the early days, gradually disappeared.

Another feature of the hydrosphere – the water on the land – was dramatically altered by settlement. In the transformation of an environment from a community of trees with numerous wet areas and abundant wildlife to orderly farmland, the wetlands lost out almost to the same degree as the forests. In some cases, the loss of wetlands was a result of the removal of the trees: without the canopy, wet areas were exposed to the full effects of evaporation. If one compares the surveyor's description of a township early in the nineteenth century with the same area today, the differences are often overwhelming. Quite apart from apparent changes in the soils, the original trees are virtually all gone, along with most of the wildlife, and the wetlands too have disappeared into drainage ditches or underground drains. There is a secret surface, about four feet below what we see, in the farmscape of Ontario, composed of an endlessly replicated dendritic pattern of tile drains that lead moisture into ditches or streams. These drains have gradually been inserted in the farm land over the past 130 years. Agricultural terrain is an impoverished environment compared to a wetland, and the extent of the loss of wetlands is now seen as a serious environmental loss. When calculating the amount of wetland lost in southern Ontario since the beginning of settlement, government scientists believe that the western quarter has lost more than 80 per cent, the eastern quarter has lost between 60 and 80 per cent, while the central half has lost roughly 50 per cent.[15]

The early settlers might have had expectations that one day their farms would be similar to those they had left behind in their home areas, but the variety and abundance of the wildlife they encountered in Ontario must have been completely alien to their experiences. This "wonder of nature" was met not so much with awe as with profligate destruction. There were three ways in which wildlife was viewed: some wild animals were considered to be pests, some were valued as food or fur, and some served as objects of recreational shooting. In all cases the wildlife was greatly reduced, and some species, such as the passenger pigeon, went from astonishing abundance to extinction in less than a century. Even by the early 1840s, a writer could say "the living, breathing denizens of the forest are various; but their numbers are fast diminishing before the destructive progress of civilization."[16]

The most notorious and troublesome wild pests were bears, wolves, and foxes. These were rapidly driven off because of their depredations on domestic livestock – indeed, a bounty was paid on a wolf scalp – and before the middle of the century they were largely restricted to the fringes of settlement and beyond. The common technique for getting rid of wolves was to trap and shoot them. Many other less notorious animals that were nevertheless at odds with farming were also trapped or hunted. Chasing and shooting a raccoon – if at night, cornering it with torches – was typically an occasion of happy excitement, sometimes for the whole family. The cougar, lynx, wildcat, marten, and other predators were killed as enemies of farm settlement, and their numbers quickly diminished around the farming areas. Even as early as 1822, in the middle of Tecumseth Township, west of Lake Simcoe, young Richard Rorke, lately from Ireland, was disappointed that wild animals were rather scarce: "during the first year of our residence there, it did not appear that any large game could be found and the tracks of bears and wolves were all somewhere else; foxes were also scarce, seldom seeing their tracks in the snow or other signs indicating their whereabouts."[17] When Rorke moved closer to the outer edge of settlement in 1848, in Collingwood Township, he only encountered a half dozen wolves in fifteen years.

The distinction between pest and pelt was not always clear. Many animals that were considered to be bothersome also had valuable fur. The main species hunted for fur were beaver, which had been pushed back beyond settlement areas by the 1840s, fox, mink, marten, otter, and muskrat. Although less popular, the skins of bear, wolf, and racoon were also used for clothing or coverings.

Certain species that were a nuisance to farming also provided valuable food to the settlers. Huge flocks of passenger pigeons, for example, could settle on a grain field and strip it clean, but they were also relished on the dinner table. Settlers in the Grand River Valley reported that, in their first few years, they would wait on the brow of the valley until the flocks of pigeons came down to settle for the night in the cedar trees. Parents would then club dozens of birds out of the air with long poles and the children would rush to throttle the birds to prevent escape. Usually only the breast meat was kept, and much of it was preserved in barrels for the winter. The deer also could be seen as a pest because of its penchant for some domestic crops, but for many early settlers, venison was a crucial part of the food supply. Deer had been greatly reduced by hunting by the 1840s, although hunting expeditions by both Euroamericans and Amerindians were still common. Another food source that was reduced in numbers – in fact, almost eliminated by the 1850s – was the wild turkey. William Smith (perhaps with tongue in cheek) claimed that "the principal *game* of Canada" was the squirrel, and that "when properly cooked [squirrels] are excellent eating, and most persons prefer them to the pigeon." He names four varieties of squirrel, and also geese, ducks, pheasants, and snowshoe hares as common wild game.[18] Seasonal restrictions on the hunting of a number of game birds were introduced in the 1840s. Many kinds of fish were also taken for food, but restrictions on taking salmon had been introduced by the 1820s. There were a number of provincial statutes related to the transformation of the natural environment: 1823, Preservation of salmon; 1830, Wolves, destruction of (c. 17); 1836, Wolves, destruction of (c. 29); 1843, Deer and game (c. 12); 1844–5, Game, to prevent improper destruction of (c. 46); 1847, Rivers and rivulets, obstruction in (c. 20, amends 7 Vic.).[19]

The custom of hunting for food or pleasure seemed to be considered a right in nineteenth-century Ontario. In 1837 the Rev. Featherstone Osler complained about hunters in the Newmarket area: "the people ... are very careless about all connected with religion. On the Sabbath ... more guns [are] heard firing than on any other day of the week."[20] Despite the obvious diminishing of some species, hunting went on in Ontario (as in most parts of the New World) as if there were no limits to the resources. That misapprehension would continue to flourish until nearly the end of the century when, for example, the passenger pigeon had already passed the threshold of extinction, and the plains bison had completely disappeared.

The thoughtless and sometimes arrogant attitude to wildlife in the rapidly evolving society that was early Ontario is best illustrated in the accounts of pleasure shooters' expeditions. One well-publicized series of letters by T.W. Magrath (which one must suspect of embellishment) gleefully describes the killing of a raccoon, in which even women participate, and the shooting of a large bear, rendering "the shaggy monster, writhing in agony." Moving on to wildlife above the ground, he makes proud calculations of fourteen or more brace of woodcock, thirty of snipe, and twenty of ducks – each in one day's hunting. Even when the shooter's accuracy was wanting, "chasing the wounded birds is esteemed good sport," and, because the harvest was usually copious, only the breast meat would be cut out and taken. Mr Magrath had a day of capital sport in December 1830 when he and his brother surprised a herd of twenty deer. Over most of the day, they pursued the scattered individuals until the Magraths had "thirteen fine deer (two of them twice hit)."[21] This kind of mindless assault on nature gives some evidence for what is known to have been a massive loss in many species and the profound unbalancing of ecological relationships by the Europeanization of the New World. The circuitous justification was that these practices made way for gainful human activities that would increase materialistic productivity – in other words, for progress.

SUPERIMPOSED GEOMETRY

The quasi-anarchy of the hunt and of the partially cleared scene was in contrast to the underlying, largely hidden order of the land survey. A bird's-eye view of Ontario would have gradually changed over the years from a pock-marked woodland, through decades of untidy cutover and burned stretches, to emerging hints of a grid pattern with large cleared expanses. The survey pattern that was laid down on the interlake peninsula, after some experimenting in the 1780s, is best described as rectangular. Originally based on the two-hundred-acre farm (although there was an early experiment with a one-hundred-acre version in the Niagara Peninsula), the norm was reduced to one hundred acres in 1815. The survey pattern provided farms of a shape generally midway between the extremely long lots that became traditional in Quebec and the square survey that appeared in one form in central Ontario as early as the 1820s. The double front survey gave rise to a "square hundred," and was extended westward to the Prairie provinces in a quarter-section (160-acre) form.[22] The roads were intended to follow the straight lines and right angles of the survey. Theoretically, the survey was to be laid out before settlement, with the road allowances and lot lines dropped like a horizontal portcullis on the landscape. This was a prime example of the attempt to impose human control on the little understood non-human matrix. Inevitably, there remained some working out of the relationship, and the relentless grid had to give way in places, sometimes permanently. Bodies of water, excessive steepness, cliffs, and surveyor errors caused straight roads to bend or halt in thousands of places across the province. And in many places the surveyed road would be replaced by a cross-country trail long used by Amerindians or early settlers, called a "given" road.

The laying of lines on the land was under the authority of the colonial government and administered by the surveyor general. This was intended to make surveying and settling efficient and orderly (and, incidentally, more sophisticated and effective than the less-controlled system across the Lakes, in the United States). External as well as internal conditions impinged on the colony, however, and the surveying was subject to stops and starts. As the demand for land and the push toward the frontier were spasmodic, so was the surveying of Ontario. The result is a survey patchwork of varying baselines, differently shaped lots and road patterns, and odd blocks and gores – ironically not very different in general appearance from the patchwork of private tracts and independent surveys in northern New York State.[23] The picture is even more complex than the kaleidoscope of survey types shown in the *Economic Atlas of Ontario* (plate 99), because parts of townships were surveyed at different times, and special needs required untypical solutions. The Euroamerican routes of communication in Ontario originated with the military plan of Lieutenant-Governor John Graves Simcoe in the 1790s, and the major corridors that developed were virtually independent of the evolving survey, though they served in places as baselines.[24] (Communications are discussed further in chapter 4, and especially in chapter 6.)

The materialization of the survey grid was gradual; even by the 1850s, many areas revealed little of its incipient geometry. One major reason for this was the great amount of land reserved by the government for later sale (Crown reserves) and for the established church (clergy reserves). Normally one-seventh of all the lots in a township were reserved for each of Crown and clergy. To the settlers, these lands quickly became irritants, being commonly uncleared and untended, so much so that the government put all outstanding Crown reserves on the market in the 1820s, by sale to the Canada Company. The clergy reserves remained a problem (as did new Crown reserves) for settlers and administrators alike for a few more decades, largely because of debates and legal wrangles over what was meant by the "established church." The buying and selling of reserves can be seen as a kind of speculation, and many ordinary farmers also set aside "reserves" of their own when they bought raw land intended for later use by their children.[25] Something closer to the modern sense of speculation took place around the major towns. Toronto, though the capital, was ringed by woodland for decades because of early grants to friends of government. On James Cane's 1842 map of Toronto, a great deal of woodland is indicated from College Street northward and Spadina westward; and J.D. Browne's map of 1850 shows extensive woods, especially to the northeast of Yorkville (Yonge St at Bloor St), and to the west on much of the land approaching the Humber Valley, surrounding Lambton and up the valley to Weston.[26]

Clearing of the trees advanced steadily so that, by the early twentieth century, many parts of Ontario, even rough land such as the Oak Ridges Moraine, had lost over 90 per cent of the original tree cover. On the moraine north and west of Toronto, Albion Township went from 40.2 to 7.8 per cent, Caledon Township from 36.8 to 8.1 per cent, and King Township from 33.6 to 6.5 per cent in woodland, between 1860 and 1910.[27] Although there were warnings of overclearing in the 1860s – the president of the Fruit Growers' Association called it an evil[28] – and a map of "waste lands" was prepared in 1908,[29] it was nearly the middle of the twentieth century before the repercussions of the massive settlement of Ontario were put into perspective and people began to recognize that the natural conditions were in crisis. A plea was made in 1942 to view nature as a vast interconnected system (today's "ecosystem") that could not continue to be carelessly damaged.[30] The establishment of a New World colony and the provision of new homes and opportunities for hundreds of thousands of people had been far from a benign creation: this human population had been a profoundly effective agent of change.

3 Agents of Transformation: An Expanding Population

> The Canadians are neither British nor American ... They are more American than they believe themselves to be, or would like to be considered ... The worst feature of the population here is one necessarily characteristic of a new country, namely, their love of money. There is such a field for making money.
>
> John Robert Godley, ca 1843[1]

LOYALISTS, REFUGEES, PIONEERS

The first Euroamerican farm settlers to enter Ontario were seeking a safe haven from a war that had not gone in their favour. These refugees from the assaults of the Revolutionary War mounted an assault of their own – the typical New World process of clearing the woodland. The population of Ontario, which grew from the arrival of the first Loyalists in 1780, did not have the supposed frontier "take-off," but was modulated through the years by various external and internal influences, sometimes growing, sometimes stagnating. A vigorous population increase in the first decade or so, when the Loyalists were getting established, gradually gave way to a depressed growth rate, reaching its nadir during the War of 1812. Renewed expansion during the 1820s and early 1830s, stimulated by the beginnings of major overseas immigration, was followed by a slump, and then a spectacular peak at the end of the 1840s when Irish immigrants flooded in. Figure 3.1 illustrates the rate of growth; even a slight levelling of the line indicates a significant slowing (as seen in the years immediately following 1812 and 1836, when the province experienced armed turmoil). Although the censuses had some serious deficiencies, there is no question that there was considerable overall growth. One demographer has calculated that the population of Upper Canada between 1837 and 1846 was increasing by 66.9 per thousand annually from births alone.[2] In fact in most years beginning before the War of 1812, the largest addition to the population, the immigrant influx notwithstanding, came from births.

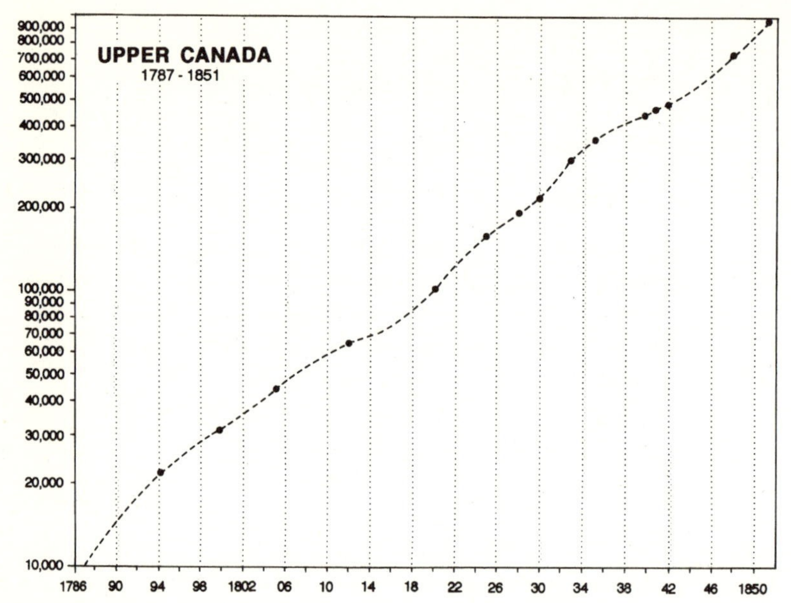

Figure 3.1 Upper Canada population, 1787–1851 (fig. 4 in Wood, "Population change")

The spreading of the population was also somewhat spasmodic, beginning at three gateways (the St Lawrence Valley, the Niagara Peninsula, and the Detroit River) in the 1780s, then growing vigorously around the new capital at York (Toronto) after 1793, then expanding generally inland away from the St Lawrence River and the Great Lakes in the 1820s and 1830s, and finally foundering at the limits of good land along the Canadian Shield edge and at the Bruce Peninsula by the 1850s. The opening of frontiers in Ontario from 1780 to 1850 was a gradual, stop-and-go process.

The population that was moving into Ontario in the 1780s was predominantly a population of frontier farmers or aspiring farmers. In addition they were refugees – granted the label and privileges of "United Empire Loyalists" – escaping the hostilities of the American Revolution and the rough justice of the postwar settling of scores. This upheaval brought a disparate mixture of people from a number of different source areas and conditions in the former thirteen colonies. In Ontario in the 1780s (at that time still officially a part of Quebec) that uprooted population gradually reorganized itself and began functioning as a new society, not exactly the same as that of its source areas or the old country. When Ontario's population began expanding beyond the original gateways early in the nineteenth century, it progressed, as did all successful agricultural frontiers, through what might be called a "fertility

Table 3.1
Pioneer demographic transition

Years from First Settlement	*Demographic Composition*	*Status of Settlement*
0	Mix of single men and young married couples Few children; men to women, ca 140–200:100 Family & household mean ca 4–4.5	Much land unoccupied
12	Marriage more common than bachelorhood; growing young families Children (under 16) account for ca 55–60% of population; men to women ca 120–140:100 mean family ca 5–5.5; mean household ca 6	Majority of farm lots occupied; clearing and planting proceeding
24	Many early families maturing Children account for ca 51–54% of population; men to women 100–120:100 mean family ca 6–6.5; mean household ca 6.5–7	Most land in local ownership; at least 1/4 cleared for crops
36	Family dispersal: most offspring leave area, but a few inherit or locate nearby Children account for 50% or less of population; men to women 98–110:100 mean family ca 5.5–6; mean household ca 6.5	Influential local families now entrenched in society and politics; land over 1/3 cleared for crops
48	Large majority of population locally born; both up and down fluctuations subdued, and total likely decreasing Children account for ca 45% of population; more women than men; mean family ca 5.5; mean household ca 6.5	All good land in use for crops or woodlots; consolidation and expansion of farms underway

transition." This is normally characterized by an overall rapid growth, based on high natural increase augmented by a sizable influx of young immigrants, followed by a steady decline in birth rate and family size (in the second half of the century, for Ontario).[3] In Ontario, individual townships might vary from the pattern, but the provincial population (as in figure 3.1) fitted this model.[4] The fertility transition in an average township would generally have followed the chronology and conditions laid out in table 3.1. It is not claimed that this table is valid for every situation, but it probably comes close to the truth for most woodland frontiers in northern North America, beginning with the vanguard of farming settlers.

CHARACTERISTICS OF A PIONEER POPULATION

The period of the War of 1812 (effectively 1812 to 1814) was a watershed for the Ontario population. Before the war the population was overwhelmingly North American by birth; after the war many townships, especially in areas recently opened to settlement, were dominated by immigrants from the British Isles. Pre-war townships were either part of the Loyalist heartland, first occupied during the 1780s, or they were "daughter colonies" made up largely of the Loyalists' offspring, an example of what Erik Bylund has described as "clone colonization."[5] The demographic character of the new pre-war townships varied, although usually it was single men or young couples from the Loyalist heartland who moved in first; sometimes there were larger families.

Information on the population of many of the new townships during the period before the first multifaceted census of 1842 can be extracted from the annual reports submitted by each township's clerk. The clerk's report to the lieutenant-governor gave a simple demographic description of each household, identified by the name of the head of the family. These reports allow one to graph the number and composition of the households and thereby get a picture of the occupants of a typical pioneer township. For example, Percy Township was a brand new frontier area in 1802, located on the south side of the Trent River about sixty kilometres north and west of the main Loyalist settlements. In 1803, when Percy had a population just over one hundred and a total of twenty-three households, there was a slight concentration of households in the ranges of two to three and six to seven persons. In 1808, with thirty-one households and a population of 175, there was a modest clustering in the ranges of one to two and five to six persons.

In 1801, when the townships of Bastard, Kitley, and the rear portions of Leeds and Lansdowne were newly opened, they contained 114 households, almost all of which were within the range of two to eight members, with the mode (twenty-two) being four. The mean household size was 5.2, and mean family size was a shade under five. By 1805, Bastard Township alone had seventy-seven households, mainly within the range of two to eight members, with a small bulge from two to five. (This township was in the second tier north of the St Lawrence River and the Loyalist heartland, approximately twenty to thirty-five kilometres overland, and its northern border was the Rideau River system.) In almost all cases, households in Bastard Township consisted of a nuclear family. In the township clerk's list of names, only two households had an individual who might not have been a family member (unlike a generation later, when non-family members were common in farm households).[6] By 1805 mean household size had gone down to 4.9, probably because Bastard was still at the forefront of new settlement and had a relatively large number of households of two and three persons, which was common for new settlements, as mentioned above. Small nuclear families interspersed in the virgin woodlands, with occasional individual or pairs of young men, was the situation one would have expected on the frontier in Ontario at this time. We can visualize the typical pioneer shanty in these new spinoff townships in the first five or so years as having contained two to five people – two adults and, if a married couple, two or three children.

In the seventy years between the first Euroamerican settlements in the 1780s and the middle of the nineteenth century, Ontario went through a pre- and a postwar generation and entered a third generation that had progressed generally beyond pioneer living conditions. One characteristic that remained constant in the early decades of agricultural settlement, both before and after the war, was the surprising amount of movement by people: most of the population in a township at the beginning of a decade could be gone by the next census ten years later. Within a township's population there were distinctions that might be characterized by the terms *transients, sojourners,* and *persisters.* The transients stayed less than two years; the sojourners tended to stay five to seven years, probably farming rented land; the persisters remained in a township through the early decades and in effect became the founding families. Around the persisters swirled scores of people in various social conditions, including sizable families, who were, to all intents and purposes, just passing through.[7] After ten years of settlement (closer to fifteen following the War of 1812), roughly half the original households could still be found in a given township. One of the Loyalist townships, Augusta, retained 51 per cent of its original households over an eleven-year span (1796–1807). In Percy Township (1803–13) 49 per cent persisted; in Haldimand (1805–15) 51 per cent; and in Darlington/Clarke (1806–15) 45 per cent.[8]

The persisters continued to account for a large proportion of the households in a township, usually one-quarter to one-third, and their households were larger than the average. At the end of an eleven-year period (1796–07) in Augusta Township, the persisting households and the families of their male offspring (traced through surnames) accounted for 48 per cent of the 1,131 residents. Persisting households, including many small second-generation establishments, accounted for 56 per cent of all households. This pattern of persistence was characteristic of both fast- and slow-growing townships.[9] Persisters, many of whom were later recognized as the founding families, quickly put down roots, expanding in size and influence, whereas the more noticeable streams of the footloose transients and sojourners milled about for a time and then moved on.

These activities of land acquisition and household establishment were primarily aimed at enhancing the success of individual farm families. Taking a broad selection of samples, Gerard Bouchard argues persuasively that such family nurturing was to be found in areas of new agricultural settlement not only in anglophone Ontario and the northern United States but also in Quebec's Saguenay region. His model applies to newly opened territory where there was a copious amount of available land and a high rate of fertility. The relatively "open" conditions for land ownership led to roughly equal distribution of land to all the young males – if not the females – in a family.[10] The vanguard of settlers, referred to above as the founding families, would have had the best access to land, and would have had better opportunities than later immigrants to provide for members of the family. In this case, success bred success, allowing the founding families to buy more land as time passed.

New Demographic Conditions after the War of 1812

The dominant demographic feature during the four decades after the war was rapid overall growth based much more heavily on immigration. New settlers from the United Kingdom – the "redundant population of Britain" that Gourlay thought ideally suited to Upper Canada[11] – began to flow towards the colony by 1816. Rapid population growth has become part of the widely accepted picture of the frontier, but in a given year the actual condition in Ontario during the first generation of influx could vary considerably. Around 1820 eastern and central Ontario – generally the initial stopping places for immigrants – were already benefitting (as illustrated by figure 3.2); but, west of the head of Lake Ontario, population growth was tentative and spotty, despite the rapid opening of new townships. The far southwestern sector continued to be generally beyond the reach of overseas immigrants and showed little population growth; it experienced variable growth in later decades, partly due to expanses of poorly drained land east of the St Clair River and Lake.[12] Other pockets of stagnation, toward the centre of the province, seemed to occur in townships dominated by sandy moraines or thinly covered limestone plains.

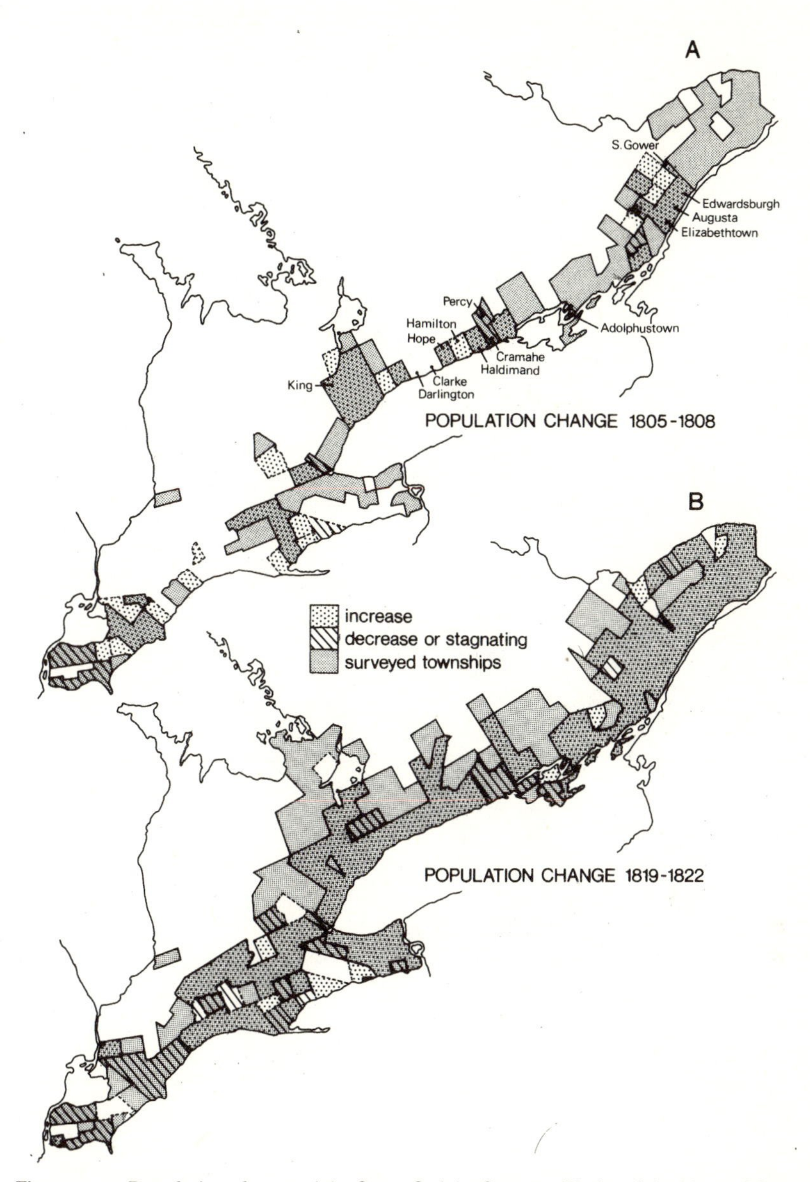

Figure 3.2 Population change: (A) 1805–08; (B) 1819–22 (National Archives of Canada [hereafter NA], RG 5, B 26; and Archives of Ontario [hereafter AO], RG 21). Increase was reckoned as 3 per cent or greater growth, stagnation as 0–3 per cent growth. The earlier fragmented population of 1805–08 is included to allow an additional comparison of the variable picture of population change on an agricultural frontier.

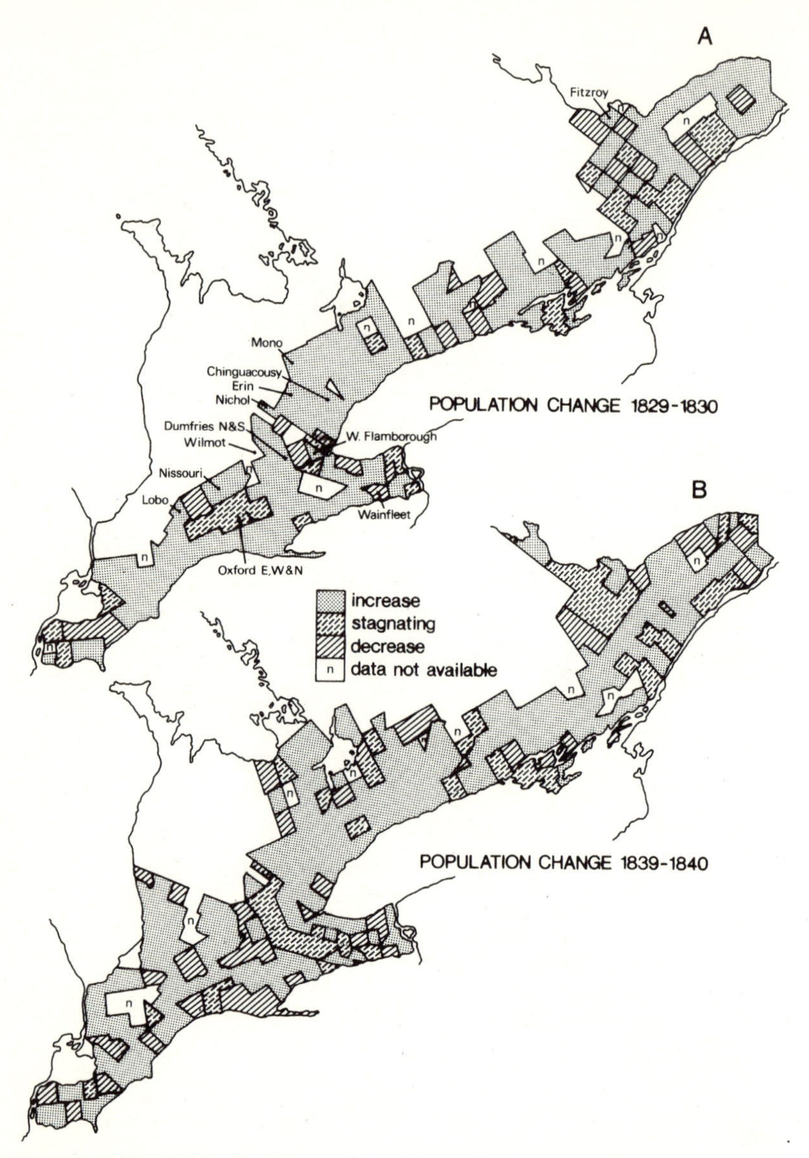

Figure 3.3 Population change: (A) 1829–30; (B) 1839–40; (C) heartland vs. fringe, 1840 (NA, RG 5, B 26; Canada, *Journals of the Legislative Assembly*, 1841). To fill gaps in 1840 census returns for townships north and northwest of Lake Simcoe and in Talbot District (i.e., Long Point and adjacent mainland), data were extrapolated based on 1839 and 1841.

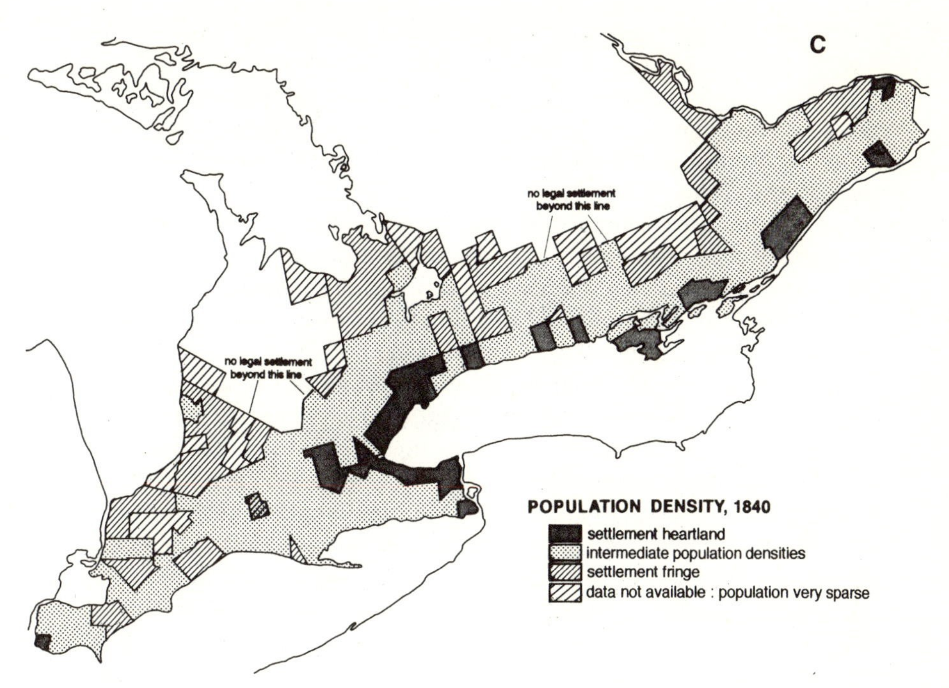

Figure 3.3 (*continued*)

By the 1820s, expansion was taking place in many parts of the province, as revealed in figure 3.3A. There was a reinforcement and spreading of the booming sector of the Home District and the adjacent portion of the Gore District. The Grand River Valley was being filled up, as was much of the rest of the north shore of Lake Erie. But there was continuing stagnation in many parts of the far southwest and in the middle Thames River Valley, while newly opened townships just to the north were growing vigorously. Long-settled townships in Niagara Peninsula and around the Head of the Lake actually lost population in 1829–30. Even some of the townships eastward on the shore of Lake Ontario lost population, after a period of rapid growth. The most striking change, however, was in the far eastern quarter, where settlers were abandoning the edges of the Shield and the very mixed-quality townships with poor drainage and stony and shallow soils.

The picture of population growth and stagnation in 1839–40, on the eve of the union of Upper and Lower Canada, presents a marked patchwork (figure 3.3B). Although expansion was still taking place at the edges of the province – that is, toward the Lake Huron shore (especially in the Huron Tract townships), north of Lake Simcoe, and north and west of Kingston – there were also losses in some townships, even near the edges of settlement.[13] Regional differentiation in the province was beginning to crystallize into the pattern that was to persist in its essentials through the twentieth century. Patterns of perennially slow growth or loss of population in areas such as the limestone flanks of the Shield, interior parts of the eastern end of the province, the Bay of Quinte, and the southern littoral of Lake St Clair, had taken shape. The northwestern shore of Lake Ontario, with its tributary back country, was displaying the kind of persistent growth that eventually led to the heavy concentration of population dubbed the Golden Horseshoe. By 1840 all the townships in the Home District that fronted on Lake Ontario, as well as the tier behind them, had a population density of over thirty per square mile.[14] On the other hand, there was a lightly populated "pioneer fringe," with a density of less than six persons per square mile, stretching discontinuously up the Ottawa Valley, along the edge of the Shield north and northwest of Kingston, around the north end of Lake Simcoe and adjacent to Georgian Bay, through most of the Huron Tract, and along the St Clair River and south to the mouth of the Thames River. (See further discussion of the extremes and their implications for social facilities in chapter 4.)

Within the provincial context of overall population increase, townships that had been growing briskly at the beginning of the nineteenth century were reaching or had reached their maximum numbers. Haldimand Township, in a desirable location on the north shore of Lake Ontario, recorded a loss of population at the end of the 1830s, as did a number of well-settled rural townships in various parts of the province. By comparison, Haldimand's near neighbour, Hope Township, reveals a general increase in the period 1819 to 1839, but with population loss in some years. Along that shore only Hamilton Township enjoyed an unabated increase. The oldest townships, settled by Loyalists on the upper St Lawrence River – represented by Elizabethtown, Augusta, Edwardsburgh – continued a trend of modest growth with periodic setbacks that had characterized their demography almost from the beginning. Many of the townships away from the waterways had had intermittent growth from the opening of settlement, usually because of a natural environment that provided difficulties for agriculture. South Gower increased to 700, but then fell back. Wainfleet Township was settled near the beginning of the century, but remained hindered by the poor drainage on the dip slope of the Niagara Escarpment. After 1826, King Township appears to have made up for its inland location on rough morainic land through the advantage of fronting on Yonge Street.[15]

Heartland and Hinterland

The distribution of population in Canada West might be described as logical, being heaviest near the St Lawrence River and the shores of Lakes Ontario and Erie, which one might expect to be the natural routes of entry, and diminishing towards the north. In reality, the population distribution in the 1840s was an unexpected product of a dynamic process that had begun fifty years earlier as three separate incursions: at the St Lawrence River entry point, the Niagara "frontier," and the Detroit River extremity. Instead of continuing to spread from these three gateways, the population rather quickly filled the intervening areas, but not in a continuous spread. There were certain places where population tended to cluster – such as around London, in the middle Grand River Valley, around the western end of Lake Ontario, and especially around the capital at York (Toronto) – and other places that remained lightly populated, such as the western end of Lake Erie and the upper portion of the St Lawrence River (around the Frontenac Axis). Towards the Canadian Shield and the Bruce Peninsula, where frontier conditions still prevailed, the population diminished.

The extremes in the distribution of the population ranged from sparse, in newly opened frontiers, to populous, in well-established agricultural landscapes and urban nodes. Within its narrow territory between the Great Lakes and the Shield, Canada West in 1840 ran the gamut of settlement that in the United States spanned hundreds of kilometres from the Ohio Valley to beyond the Mississippi in the west.[16] In modern terminology this range can be described as running from "heartland" to "hinterland," or, in the terms of a study of population growth rates in two newly developing areas, from "core" to "frontier" through "intervening space" (figure 3.3C).[17] There were a number of discontinuous areas of heartland, or core areas (indicated by the darkest shading on figure 3.3C), but a relatively continuous outward-pushing belt of fringe, or frontier, settlement. By 1820 it was not unusual for there to be a few squatters ahead of the frontier, or even beyond the survey.

The values for the two extremes, heartland and hinterland, are taken from the classic debate stimulated by the superintendent of the United States census in 1890, and effectively elaborated by F.J. Turner. The beginning of farm settlement (as opposed to lumbering or hunting) was represented by six persons per square mile.[18] For Canada West this could be translated as the equivalent of one family per square mile (640 acres) and referred to as the frontier, or hinterland (shown as the outer cross-hatched edge on figure 3.3C). At the other extreme was a density of six of more families (thirty-six or more persons, even fifty in a few townships) per square mile, representing the heartland, the densest settlement.[19]

Between heartland and hinterland was territory occupied by people engrossed in the process of developing a farming livelihood (the stippled area on figure 3.3c), with population densities between six and thirty-six persons per square mile and an intermediate level of social structuring and related facilities. (See chapter 4.) During the 1840s the heartland grew vigorously. Measured in terms of a township population density of thirty-six or more persons per square mile, there were 120 heartland townships in 1852, whereas in 1840 there had been thirty-seven.

Sex Ratios, Age Composition, and Transience

As we have seen, the growth and distribution of the population varied from township to township. It is now appropriate to consider the other fundamental characteristic of the demography, its composition. The ratio of men to women – a critical aspect of a pioneer population – gradually moved toward equilibrium as time passed. By the end of the 1830s, when the ratio of men to women for the province as a whole was 110:100, ratios for most long-settled townships were actually lower than that, with nearly even numbers of men and women. There were exceptions, however, such as the Loyalist township Elizabethtown, which suddenly jumped to 131:100, and Hamilton Township, which jumped to 153:100. One could speculate on the reasons for such sudden rises; one possibility, of course, could be inaccuracies in the existing data. In these two cases the more likely reason was that rapidly growing towns (Brockville and Cobourg, respectively) provided conditions for periodic fluctuations. In the relatively small populations characteristic of most townships – generally numbering only in the hundreds and rarely over 2,500 through this period – one or two dozen transients could have a major effect. In a few fringe townships, the coming and going of men engaged in the timber industry could drastically modify the number of adult males; but, as is demonstrated below, men were not the only transients.

It is possible to calculate provincewide ratios of men to women during the 1830s and 1840s. Whereas in 1835 there were 117 men (age 16 or older) to 100 women, by 1841 the ratio was 110:100. In 1835 the ratios ranged widely from pioneer Huron District, at 161:100, to long-settled Glengarry District, at 107:100. In 1841 Huron stood at 127:100, and Glengarry at 103:100. In 1851, after the heavy Irish immigration of the late 1840s, there were 115 men (now 15 or older in census categories) to 100 women in the province. The Huron County ratio was 120:100, and in Glengarry County, 99:100.

The under-sixteen, or children's, proportion of the population was a distinctive part of the demography of early Ontario. (The division at sixteen is less than ideal, because many children were employed before that age, but we are bound by the categories under which the township clerks were instructed to report the population, beginning in the 1790s. In the 1842 census a more realistic division was used, with the adult categories beginning with age 14; this continued in the 1848 census. In the 1851–52 census, adulthood was taken to begin at age 15.) From the first two or three years of settlement, the number of children began to rise rapidly. In the townships settled through the 1820s to the 1840s, the peak proportion of children was usually reached in twelve to fifteen years, at which time children accounted for 52 to 55 per cent of the population. The distinction needs to be made that, although the frontier, with its demands for strength and aggressiveness, may have been a men's domain, it rather quickly became a children's domain if the land was conducive to agricultural (and therefore family) development. A significant number of women had their first child by the age of sixteen. This was part of the "fertility explosion" before land began to get scarce and the younger generation began to consider leaving the farm.[20]

The importance of children and women in the changes in the population demonstrates that mobile men were not always the chief demographic influence. In fact the fluctuations in the numbers of people were not primarily due to the comings and goings of unattached men, although they were an important mobile group. An analysis of available long population series, some beginning before the War of 1812, reveals that, although some fluctuations may have been caused by migrant men, just as many were caused by changes in the number of children, or women and children, or by all components combined. Among the postwar townships, only a few appear to have been affected specifically by the movement of men. In Fitzroy, near the heavy timber exploitation, the increase of 1829–30 and the decrease of 1837–38 are examples (see figure 3.5). But the increases of 1827–28, 1832–33, and 1836–37 were related as much to the proportion of children as to that of adult males, as was the decrease of 1828–29. In Chinguacousy and Nissouri none of the changes can be attributed primarily to the men. The exodus of migrating men was probably the reason for the slowing of growth in the population of Lobo Township, on the western frontier, in 1832–33 and 1838–39; but apart from those years, any changes were broadly based. The calculations for townships with small populations could be skewed by the arrival or disappearance of a few families, and even more populous areas could be disrupted by the opening of a large block of land to settlers, or by the establishment of a new industry, such as a foundry, or by canal construction.

As indicated earlier, transiency was a ubiquitous feature of the early agricultural population in Ontario. It is possible to take a closer look at this feature through evidence in annual reports on population by household from township clerks.[21] An example of a township that had shown intermittent growth, in which transiency was important at certain times but not at others, is provided by South Gower (see figure 3.4). It was a township that reached the maximum occupancy of its roughly one hundred farm lots shortly after the War of 1812, having been opened to settlement about the turn of the century. Of the forty-two households in South Gower in 1813, fifteen (36 per cent) were still there in 1825; some, in addition, appear to have been succeeded by offspring. This was a slightly lower persistence rate than in many of the early-settled townships, suggesting fewer opportunities for success in South Gower.

Moving West and North: Erin Township

In the 1820s many new townships were being opened to the west. "The west" for Ontario in the three decades leading up to the middle of the century was a gradually expanding belt of newly surveyed townships reaching north of the Oak Ridges Moraine toward Georgian Bay, northwest of the head of Lake Ontario toward the Bruce Peninsula, and west of the Grand and Thames River Valleys toward Lake Huron. At 1840 this belt of frontier settlement stands out on the map of population density as a lightly populated fringe in contrast to the populous heartland along the Lake Ontario shore (*see* figure 3.3c).

One might expect these new western townships to be an enticement to residents of a township of variable land quality like South Gower. Our expectations are not borne out in this case by the comings and goings of the people of this township during the period 1825 to 1830. Not only did the population increase – albeit haltingly – during this period, but most of the households persisted. In 1827, 78 per cent of the 1825 households were still resident, and in 1830 70 per cent remained. This does not mean that the demographic scene was stable, because the list of households reveals that a number of people left the township and returned two or three years later. The most notable permanent loss was the Mahlon Beach household. Beach was a township

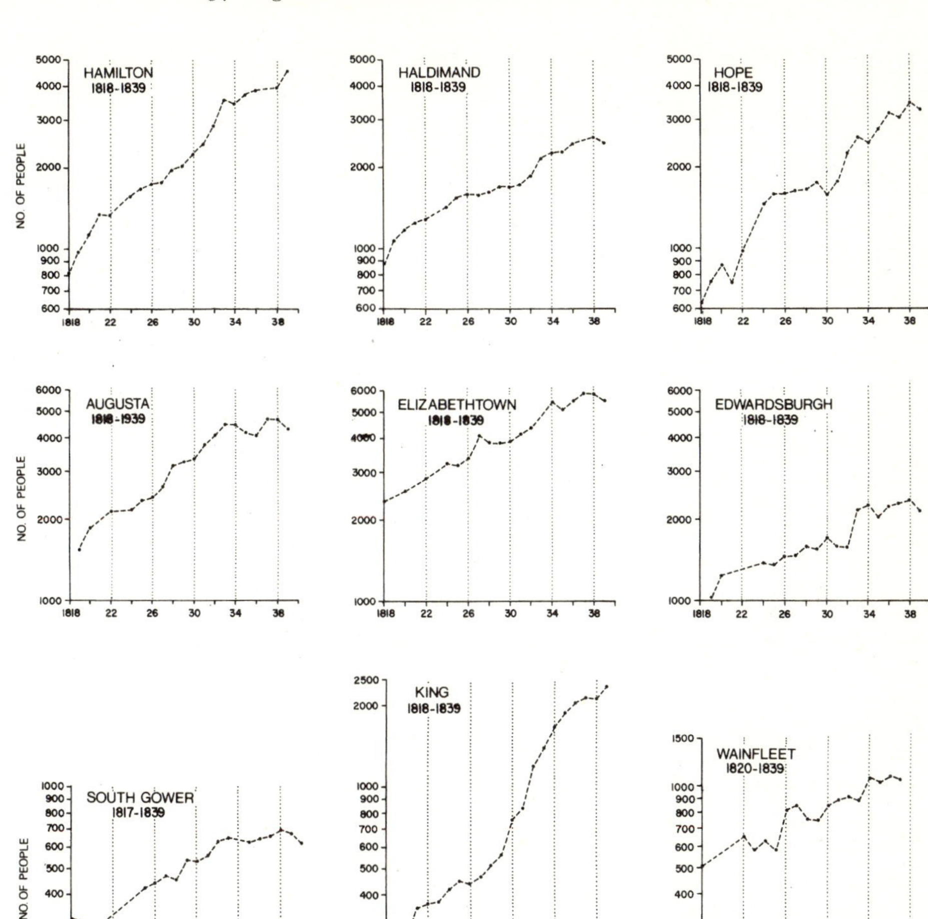

Figure 3.4 Population, 1817–39, of selected townships opened for settlement before 1810 (NA, RG 5, B 26; AO, RG 21)

assessor in 1825 and 1827, when his household totalled eleven, but he had left the township by 1828, following an earlier departure by one of the other half-dozen Beach families. South Gower had the characteristic fluidity of the population at the beginning of the nineteenth century, but, at least for the five years ending in 1830, it was not being depopulated by a rush to the west, or anywhere else.

An example of the townships newly opened for settlement further west is Erin Township in the Gore District. Part of the 648,000-acre cession by the Mississauga Amerindians in 1818 (later Peel and Halton Counties), Erin Township was surveyed into lots in 1821. In natural

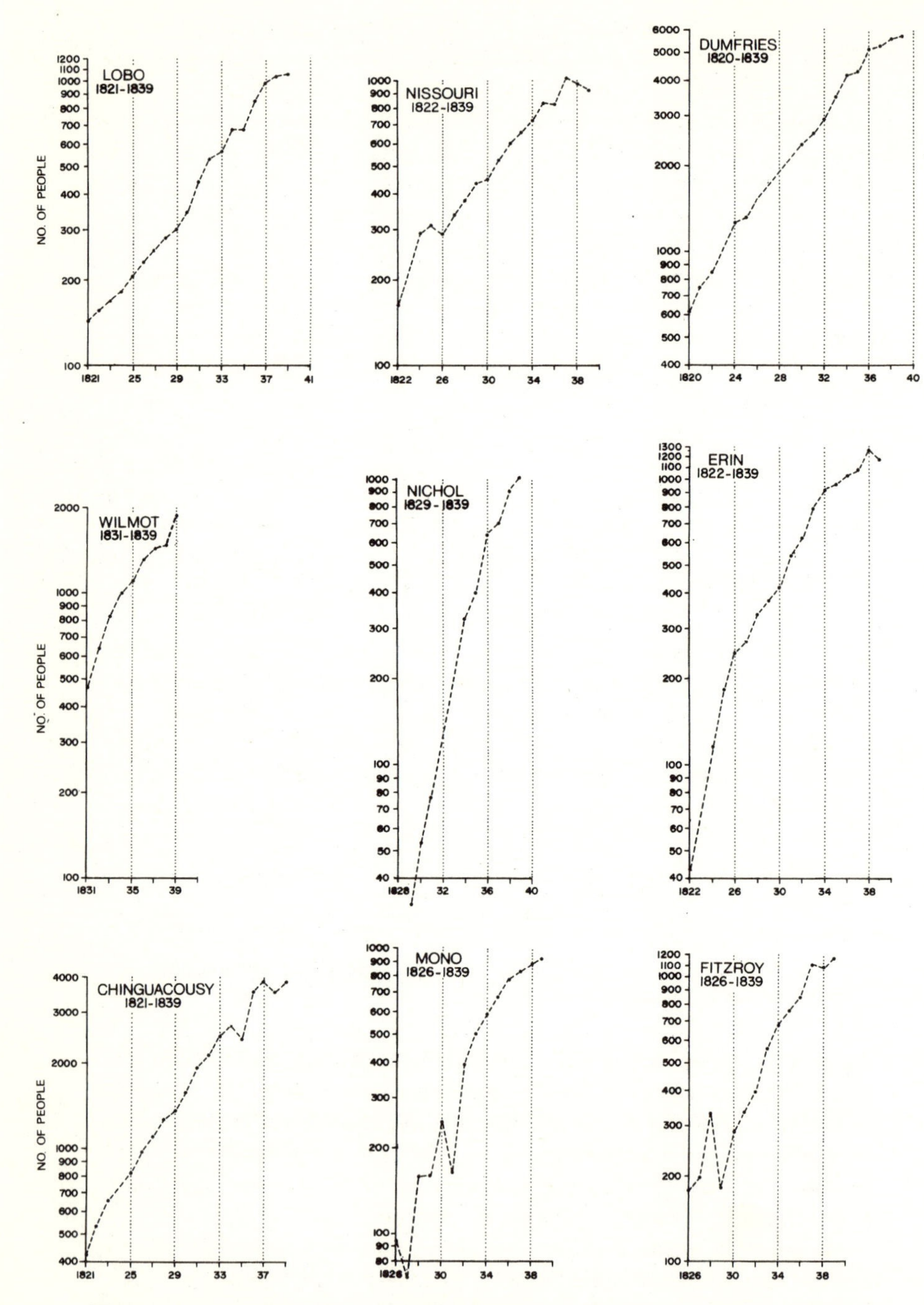

Figure 3.5 Population, 1820–39, of selected townships opened for settlement after the War of 1812 (Canada, *Census of 1870–71*, vol. 4)

endowments for agriculture, it was closest to some townships settled earlier that had suffered fluctuations in growth (South Gower, Edwardsburgh, and Wainfleet, shown in figure 3.4). Erin lay on top of the Niagara Escarpment and thereby was meagerly blessed, having shallow soils or heavier morainic deposits of mediocre fertility, both interspersed with wet patches. Unlike other townships with poor soils, however, the population growth seemed to be unimpeded by the natural deficiencies during the first fifteen years (see figure 3.5). In 1822 there were 43 souls; by 1841, 1,200.

All the new western townships had a similar exponential growth – shown most dramatically by Erin, Nichol, and Wilmot – such as had occurred in the pre-war period in only a few well-endowed townships. Even townships with somewhat precarious economies regained the momentum; for example, very hilly Mono recovered from an initial loss of settlers, and timber-based Fitzroy, in the Ottawa Valley, overcame early losses of nomadic loggers. The main advantage of postwar over pre-war settlements was the strong influx of overseas immigrants who pushed into most areas, but especially the new townships, in their eagerness to make a livelihood, even in conditions with which they were unfamiliar. Typically, the rapid growth would continue for a number of years until the farmers came to terms with the environmental limitations of a given township. In almost all the townships the growth levelled off or receded after fifteen years.

Notwithstanding its rapid early growth, Erin shared important demographic aspects with long-settled townships. Its original households tended to persist: 52 per cent of the twenty-nine households of 1824 were still there in 1834. The loss of households over this period was gradual: after two years (in 1826) 72 per cent were persisting; after seven years (1831) 62 per cent were. Over the ten years, five of the households disappeared and then returned. Many households also appear to have had members coming and going from year to year.[22]

In important respects Erin Township also differed from older townships. The founding families were inundated by the flood of newcomers following on their heels. Although over half of the 1824 households were still in the township in 1834, at the latter date they only accounted for 9 per cent of all households (unlike the 25 to 35 per cent in the pre-war townships). The persisters, along with their relatives or offspring (based on surname), only made up 17 per cent of the 1834 population of 859. The founding families did uphold the persisters' reputation of having larger than average households – 7.1 compared with the township average of 5.3.

Again, Erin differed from most parts of the province in that the proportion of men to women gradually increased from 1825 to 1839, beginning with an adult male-female ratio of 107:100 in 1825 (population 185), and rising irregularly to 123:100 in 1834 (population 948) and 128:100 in 1839 (then falling to 110:100 in 1840). In one other respect Erin Township was unusual. Many of the population graphs show a setback between 1837 and 1838.[23] The Rebellion of 1837 appears to have had more effect on the population in some parts of the province than in others, as did the War of 1812 a generation earlier. In most cases, townships lost men at such times of unrest, as did Erin, but at the same time, Erin reported an increase in population brought about primarily by a rise in the number of children.

THE 1840S: THE MODERN CENSUS ARRIVES

By the 1840s a broader range of demographic information begins to appear from various attempts at a full-spectrum census. From 1842 on, the census began to gather information on place of birth, religious affiliation, occupation, age, marital status, crops, and dwellings, as part of a 120-column enumeration. Although these early attempts are far from satisfactory in completeness, reliability, and survival of material, some gross patterns begin to take shape.

In 1842, when the population was 487,053, the largest foreign-born group was Irish (with 16 per cent), followed by English (12 per cent), Scottish (8 per cent), American (6.7 per cent), and European (1.4 per cent). The remaining 56 per cent of the people were born in Canada (see figure 3.6). The mean household size, based on the number of dwellings, was 6.2. In religious affiliation, four main denominations accounted for 75 per cent of the population – the Church of England for 22.1 per cent; Presbyterian, 19.9 per cent; Methodist, 17 per cent; and Roman Catholic, 13.4 per cent.[24]

By 1848, when the total population was 725,879, the mean household size was 6.45 persons. Religious affiliation had changed only slightly for the Church of England (22.9 per cent), and for the Presbyterians (20.4 per cent); but growth was substantial for the Methodists (19 per cent), and for the Roman Catholic Church (16.4 per cent). At the end of 1851 the total population was just over 952,000.[25] The mean household size had gone up to 6.5 persons, but the proportion of children had gone down, 44.8 per cent of the population being under age fifteen. Of the immigrant population, 18.5 per cent were born in Ireland, 8.7 per cent in England, 8 per cent in Scotland, 4.6 per cent in the United States, and 1.2 per cent in Europe. Nearly 60 per cent of the population was now native born. The four main religious

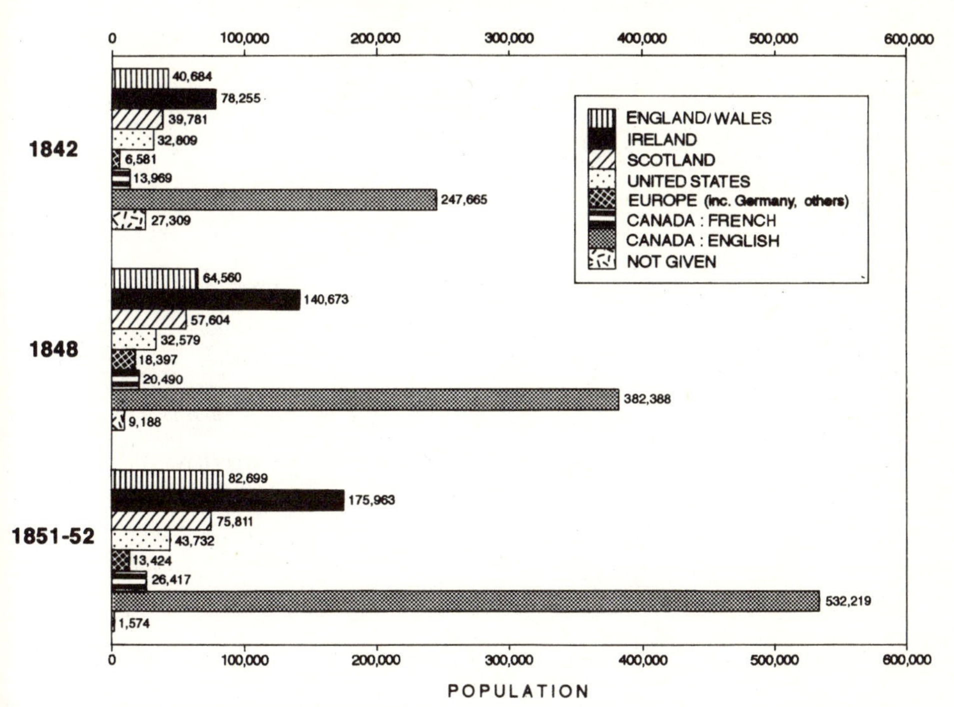

Figure 3.6 Birthplaces of the Ontario population: 1842, 1848, 1851–52 (Canada, *Census of 1870–71*, vol. 4)

groups accounted for 85 per cent of the population. Each of the four had made gains, with the Methodists gaining over 3 per cent. The Church of England was still the largest denomination, at 23.4 per cent, followed by the Methodist (22.4 per cent), Presbyterian (21.4 per cent), and Roman Catholic (17.6 per cent) churches.

The demographic scene that materialized in the three or four decades leading up to the building of railways around 1850 had the dynamics common to many New World agricultural frontiers, particularly what are called midlatitude settler colonies. In summing up the demographic evolution of Ontario before the 1850s, perhaps the most fundamental element was the ratio of women to men. It was not unusual for there to be twice as many men as women in the first few years of rural settlement, and unmarried women were rare. This ratio gradually modulated over time, especially as children moved into adult age, so that after twenty years there would have been less than 120 men to 100 women (see table 3.1 above). Another central feature was the rapid increase in the number of children (this num-

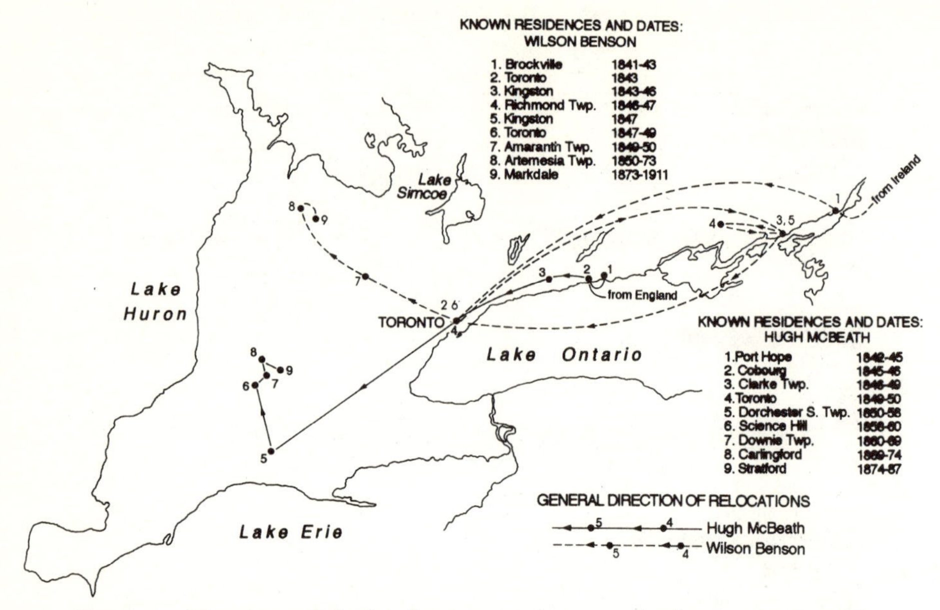

Figure 3.7 Migration paths in Ontario of two immigrants, early 1840s to turn of twentieth century (Katz, *People of Hamilton*, 94–111; Houston and Smyth, *Irish Emigration*, 152–5; unpublished McBeath family papers)

ber was to gradually level off in the second half of the century). The high birth rate had important repercussions for the overall growth of the rural population and the filling up of the land, as well as for matters of public policy, such as the provision of education. The third central feature was the great mobility of the people: a readiness to move was a dominant characteristic of the population throughout the period under study. Although popular literature made much of the allure of the West, an observer would be hard pressed to find a common orientation in the odysseys of Ontario migrants. The patterns of two settlers, Hugh McBeath and Wilson Benson, are shown in figure 3.7.

The illustrated transit of Hugh McBeath across Ontario reveals no invariable preference for a westward orientation; although he did move west five times, he then moved to four locations east and north. Although moving seemed to hold no fear for nineteenth-century residents of Ontario, the search for opportunities would normally have remained within the district unless relatives or friends were beckoning from further afield.[26] McBeath was representative of the mid-nineteenth century migrant. Starting from his home in John O'Groats, at the northern tip of Scotland, he went to England in search of employment. A few years later McBeath decided that opportunities were better in Ontario, and he settled in Port Hope, after disembarking at Cobourg. In a series of short stays, he set down almost equally in urban and rural locations, and took up different occupations (including farming from 1850 to 1869, followed by shopkeeping).[27]

Wilson Benson's trek on the tableau of mid-nineteenth century Ontario begins about the same time as McBeath's. Benson had been in pursuit of employment in various parts of Ireland and Scotland before marrying and coming to Canada. If anything, he was more mobile than McBeath, mainly because of his seasonal employment on lake vessels during the 1840s. He retraced his steps on a number of occasions and was employed in a great range of occupations until, like McBeath, he eventually chose land for farming near the outer fringe of settlement. With advancing years and infirmity, he too turned to shopkeeping in a nearby village.[28]

The pattern of demographic change was repeated throughout the half century or so that landtaking proceeded across the province. From township to township the first settlers would entrench their influence; the numbers of men to women would slowly diminish, at least in purely rural townships (in the large towns the numbers of women came to equal or even surpass that of men at a relatively early date); growth of population would continue (but probably with periodic setbacks), with the children (under sixteen) coming to account for over half the population; and a shifting crowd totalling more than the number of permanent residents would pass through, pausing for months or even a few years, during each five- to ten-year span. The first generation set the demographic model that the next generation in a township would repeat in a muted version, at least into the second half of the nineteenth century. Eventually transience became more of an exodus than a reciprocal flow in and out.

Intertwined with transience was what came to be called the rural-urban shift. Some rural areas reached maximum population in the 1840s and, from that decade on, reported a gradually diminishing total, at least in the nineteenth century. It became clear that a major component in this shift was young women abandoning rural areas. There was little in the way of employment for a young woman in the country; she could become the wife of a farmer or of someone in an agriculture-related occupation. Carpenters, masons, and equipment-makers were needed for the rudimentary farms. As a township filled up and agriculture passed beyond its initial foundation phase, occupational opportunities levelled off or disappeared. Young men, but

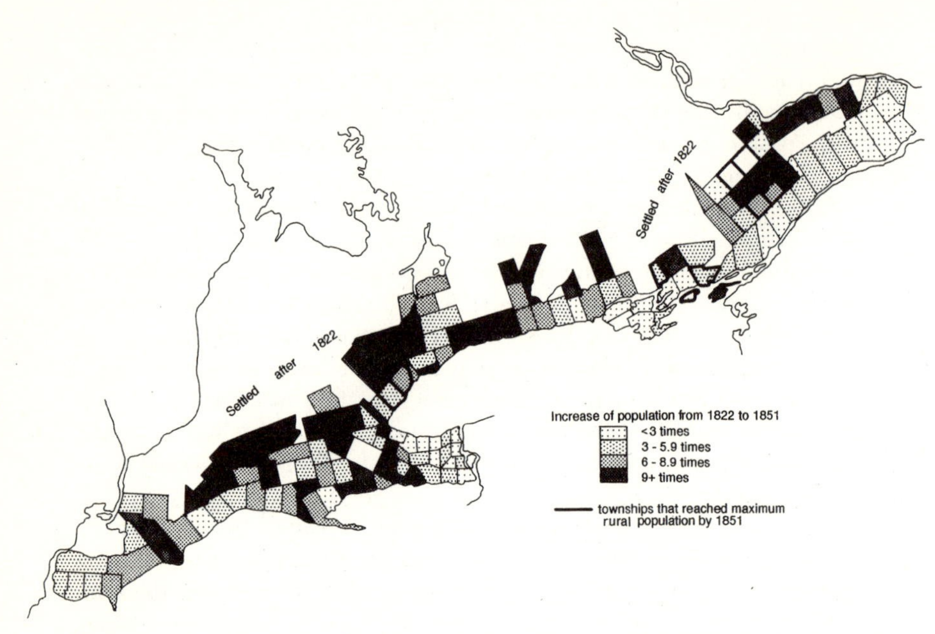

Figure 3.8 Population growth, 1822–51, in townships opened for settlement by 1822 (NA, RG 5, B 26; AO, RG 21; Canada, Board of Registration and Statistics, *Census of 1851–2*). Some 1822 figures have been extrapolated from years before and after.

young women even more so, found that they had to move to find employment.[29] They joined the stream that was proceeding rapidly from rural Ontario to urban places by the 1830s. Whereas most rural areas still had markedly more men than women, the largest urban centres started to have more women than men in the 1830s. The ratio of men to women (age sixteen and over) in Kingston in 1835 was about 93:100, and Toronto had almost equal numbers of each.[30]

A visual summation of where the most dynamic rural population change was occurring in the generation after the War of 1812 is seen in figure 3.8. All the townships were more populous in 1851 than they had been in 1822, but many older areas (such as the triangle between the St Lawrence and Ottawa Rivers, and in the Niagara Peninsula) had grown relatively little. A number of townships had already reached their maximum farm population by 1851 (or earlier). The provincial population as a whole, including that of towns, where growth was generally strong, had increased about sixfold. As one might expect, the largest proportional rural change had occurred in townships opened to settlement around 1820, but there were also comparable increases in some older areas where rich land encouraged growth.

A NEW WORLD MOSAIC

The first half century after the establishment of Upper Canada was a period of substantial population increase (figure 3.8). In the 1830s and 1840s the population generally exhibited the demographic characteristics representative of the "take-off" of the fertility transition, depicted in the previous section: overall rapid growth, a large proportion of children, a surplus of men in rural areas, a low death rate, and a great deal of transiency. By the end of the 1840s, however, distinctions were beginning to emerge: growth in rural areas was slowing down or stopping, while urban places were growing rapidly. The second half century, after the 1840s, was to be a period of reduction in immigration (indeed, emigration was usually higher) and a deceleration of the population growth, but with urban places continuing to attract new residents. In this general picture there were a number of localities distinguished by ethnic and cultural characteristics that played a part in shaping the landscape.

Even from the opening of Upper Canada, in the 1790s, the population was an unhomogenized assemblage of groups separated by origin and distance. Around the major gateway, the upper St Lawrence River, the population was primarily composed of American-born English-speaking Loyalists, with some Gaelic-speaking former soldiers at the east end and a group of Six Nations Amerindians on the Bay of Quinte. The Niagara gateway was similarly occupied by American-born English-speaking Loyalists and, in the adjacent Grand River Valley, the largest congregation of the loyal Six Nations people. The far western gateway, along the Detroit River, was occupied by a long-established French-speaking community, with nearby Amerindian groups around Lake St Clair and at places on the Detroit River and up the Thames River. In the hinterland, especially along the waterways, were other small transient native groups, perhaps two to three thousand, from Georgian Bay and the edge of the Shield south to the Great Lakes.[31]

In the first half of the nineteenth century, mainly after the War of 1812, while the population was gradually filling in the territory and linking the gateways, immigrant groups entered the province from overseas. Some of these groups were cohesive enough to be able to impose cultural characteristics that gave unique identities to certain areas. Throughout the period under study there was a continuing influx from the United States (as intimated by figure 3.6), except for a few years following the War of 1812. The Ontario scene came to be, if not an ethnic patchwork, at least generously sprinkled with colourful nodes of distinctiveness. Many of these unique settlements persisted so strongly that a later commentator could argue that, if he depicted "a farm settlement on the St Lawrence below Brockville, another of a group of farms in the German settlement of Waterloo County, ... another of the Paisley Block in Wellington, another of a French settlement in Essex, you would hardly believe that they all represented different sections of the same province, and you would admit that the origin or nationality of the people had much to do with their condition."[32] This varied demographic scene was in place by the middle of the century and, in fact, would have been more pronounced than in 1898, the year of the quote above, because many of the distinctive settlements were at the frontier or in isolated locations.

The French had been numerous along the Detroit River since the 1750s, had appeared at Penetanguishene when Drummond Island was yielded to the United States in 1825, and were moving across the Ottawa River into the far eastern end of the province from Canada East (Quebec) before 1840. Also near the eastern extremity of the province, on the St Lawrence River, were the Gaelic-speaking Scots that the prudish John Howison described, about 1818, as "blunt and uncultivated" as well as "unambitious ... dirty, ignorant, and obstinate."[33] This settlement had its origins in the demobilization of Highland Scottish soldiers from British regiments in the 1780s. There were other heavily Scottish areas in the province (commemorated here and there by the name "Scotch Block," "Scotch Line," or "Scotch Settlement"), most of which were English-speaking. The Talbot settlement along the Lake Erie shore south of London attracted a large number of Scots, many originally from the Highlands (and also apparently enticed the obnoxious traveller, Mr Howison, who came a very long distance in primitive conditions to observe and criticize another group of his struggling fellow Scots).[34]

The Irish were found throughout the province, and, up to the middle of the nineteenth century, they far outnumbered the English and the Scots. In some areas Irish-born people formed a population matrix that engulfed English, or Scottish, or even native-born Canadian families. This was true, for example, in Douro Township (on the east side of Peterborough), in a number of townships in eastern Ontario, in the northwest corner of York County, and in most of Simcoe County in the late 1840s.[35] The Irish have to be recognized as two separate groups, at least in religious affiliation – the largest one Protestant (but heavily from the south), and the other Roman Catholic. There were some unhappy areas in Ontario where the two groups faced one another uneasily and not always peaceably. For example, in southern Simcoe County, the territory of the Catholic Irish in Adjala Township was gradually ringed by Orange Lodges that sprang up in surrounding Protestant Irish areas by the 1870s.[36] There was no love lost between the hamlets of Loretto, Keenansville, and Colgan on one side, and Penville, Connor, and Hockley on the other. If one could recreate the verbal interchanges of such an area in the mid-nineteenth century, one would be assailed by the strikingly incongruous cadences of Old World dialects ringing through the rough New World landscape, with little of what one could call North American speech. One of the puzzles of Ontario history is the apparent slide of the Irish from a position of numerical dominance of those born outside the province, from the 1820s well into the second half of the century, to a rather modest proportion of the population by the First World War.[37]

The German settlements were particularly noticeable because, at least initially, they were distinguished by both language and religion. The major settlement in Waterloo County, beginning in the first decade of the nineteenth century, was composed of German-speaking Mennonites, and a smaller concentration of this group went to Vaughan Township near Toronto. A number of non-Mennonite, German families had been brought into Markham Township, just east of Vaughan, by Lt-Gov. John Graves Simcoe in the 1790s. Some of these German cores, whether religious or secular, were later extended by other German speakers who located nearby. The German Mennonites have always been noted for their successful farming, both in Ontario and in the states of Pennsylvania and New York from which they emigrated to Canada. Religion, not ethnic background, characterized the successful settlements of Quakers, who located around the Bay of Quinte, around Yonge Street south of Lake Simcoe, in Pickering and Uxbridge Townships, in the Niagara Peninsula, and in one or two smaller groupings further west. Quakers became assimilated more quickly than the foreign-language groups, especially after schisms eroded their communities in the 1820s and later.

Perhaps the most distinctive groups that were found in the farming landscape of Upper Canada were African Americans who had fled the United States. Most of the black settlements offered a land-based opportunity for economic independence along with the advantages of an amenable, supportive community. Black settlements were described figuratively as being a terminus of the Underground Railroad; that is, they provided a place of refuge for blacks who were able to cross the border – often clandestinely – from the United States. These settlements flourished mainly during the 1830s to the early 1860s, but there had been black settlers in the province from the early days of British arrival, including a number who came as slaves of Loyalists in the 1780s. By the 1820s, black families were to be found in many parts of the province, including towns such as St Catharines and Amherstburgh. According to the 1842 census, Toronto had 470 "Coloureds"

("Race Noire"), and the Home District had a total of 803 "Coloureds," with individual townships ranging from zero to sixty-seven. The 1848 census lists 516 blacks in Toronto, the majority of whom F.H. Armstrong believes lived in a newly developing area on the northwest fringes of the town, with a smaller concentration at the northeastern edge of the central business area.[38]

The main focus of black immigration after the War of 1812 was the southwestern corner of the province, although the first area designed for black group settlement appears to have been near the military road in Oro Township northwest of Lake Simcoe. Fifteen Ontario "men of colour" were granted lots in this location, probably as a reward for service in the War of 1812,[39] and a few settled there at the end of the 1820s, but within a decade or so they were surrounded by ex-military men and other settlers of British origin. The black settlements in the southwestern corner of the province were explicitly envisaged as colonies, and each had certain utopian aims, such as increasing literacy and broadening the range of gainful occupations practised by colony residents.[40] The largest colony, which had forty-five families in 1851 and a population of about 400 by the end of 1852, was Buxton, twenty miles southwest of Chatham. It was organized at the end of the period of this inquiry, in 1849.[41] A much earlier initiative was the Wilberforce colony, established in Biddulph Township, a dozen miles north of London, in 1830. This colony was established on 800 acres of land, and at its maximum numbered under 200 residents. The third colony of considerable size was located in Dawn Township, ten miles east of Dresden, in the early 1840s. This colony eventually grew to approximately 500 residents.

The census of 1851–52 reported, in its "General Digest of Origins," that "there are about 8,000 colored persons in Western Canada [i.e., Canada West]." The residents of the black colonies gradually began to find opportunities more attractive elsewhere, and the colonies diminished after a couple of decades. A significant number of black settlers returned from Ontario to the United States after the 1862 declaration of emancipation.

A small but unique component of the population was composed of a variety of Amerindian groups. By the nineteenth century they were no longer feared by settlers in Ontario, and indeed were commonly viewed as somewhat romantic figures in a picture of quintessential Canada. Writers like Catharine Parr Traill rhapsodized on her neighbours in their picturesque "wigwams," their "swarthy complexions, shaggy black hair and singular costume" highlighted by "the red and fitful glare of the wood-fire that occupied the centre of the circle." The picture was completed by "the Indians singing their hymns ... in the

Indian tongue" – but hymns from the European experience nonetheless.[42] Now and again, however, Mrs Traill and her farming neighbours would be startled by the silent appearance of an Amerindian, because in much of the back country, especially along the Trent and Simcoe/Severn drainage systems, First Nations people lived in small migratory groups.

Before 1850 about three dozen Indian reserves of greatly varying sizes were located across the province, usually on a water frontage.[43] For example, the whole Bruce Peninsula was identified as a reserve; others were located along the shore of Lake Huron, on the St Clair River, the Detroit River, the Thames River, between Lake Couchiching and Georgian Bay, on the south side of Lake Simcoe, north and southeast of Rice Lake, and on the lower St Lawrence River at Cornwall Island. The two largest reserves, with relatively sedentary, sizable populations, were the Six Nations reserves on the lower Grand River (2,223 residents in 1843) and at Tyendinaga on the Bay of Quinte (just under 400 in 1851).[44]

The population of Ontario in the first half of the nineteenth century, though including groups in notable contrast to the majority, was overwhelmingly British in origin, with a large Irish majority in the 1830s and 1840s. The native-born portion of the population accounted for well over half the total by the census of 1851–52, and to an increasing extent Ontario was becoming an amalgam of Old World (primarily British) and New World components (in its population as well as in its social evolution, economic initiatives, and technological developments). Whatever their origins, the people of Ontario at mid-century, having survived the Rebellion of 1837 and the loss of preference in British markets, were still predominantly committed to the Crown, and in many places were ostentatiously loyal. As an agent of landscape change, the population became more influential as numbers increased to one million in 1852. Under the umbrella of imported British institutions for maintaining order and managing the society as a whole, grassroots organizations ministering to more routine social needs began to evolve. The next chapter elaborates on that development.

4 Building a Social Structure

> This Colony, (which I mean to show forth with all the advantages of British Protection as a better Government than the United States can possibly obtain), should in its very Foundations provide for every Assistance that can possibly be procured for the Arts and Sciences, and for every Embellishment that hereafter may Decorate and attract Notice, and may point it out to the Neighboring States as a Superior, more happy, and more polished form of Government.
>
> John Graves Simcoe, 1791[1]

Upper Canada had difficulty living up to Lt-Gov. John Graves Simcoe's expectations. In the four years that Simcoe governed the colony, he did his best to inculcate in the ragtag society "British Customs, Manners, and Principles."[2] Most of Simcoe's successors attempted to do something similar, each according to his own priorities and formula. But, for this colony at the very edge of international trade (and ultimately at the edge of imperial preoccupations), sophistication and economic success were slow to come. The colonial administrators designed and built a settler colony, with the usual elite class of administrators, lawyers, and merchants.[3] This influential group ensured the necessities for trade and the movement of primary materials, and they set up a rudimentary legal system with an emphasis on protecting property, one "embellishment" inherited directly from the old country.[4] This framework of political control quickly became solidified and was reinforced through time. It led by the 1820s to a widely recognized irony: a society engrossed in farming was supported by an agricultural sector that had little political influence beyond the township. This is not to suggest that the influential persons did not own land, but the real "movers and shakers" of the province moved in urban, not rural, circles.[5] Apart from a few institutions imported from the old country, a social structure was slow to take shape, to the particular disappointment of ordinary settlers.

FROM REFUGE TO COLONY

Meeting Needs

The colony did not completely stand still during the first few decades. The 10,000 residents of 1787 became 35,000 by 1805 and 65,000 by 1810, but by the latter date they were spread over an area more than ten times as large. With a more rapidly increasing population after 1815, settlement activity gained momentum in most districts, and social organization and intercourse became more complex: more cases of poverty, the entry of large numbers of mobile people, an increase in crime, the appearance of epidemic diseases, and demands for more public services and a more sensitive and "modern" government. An inquisitive individual like Robert Gourlay, who in 1817 asked a series of superficially innocent questions about conditions in the colony, could become a lightning rod for dissatisfaction and rouse the wrath of the governing elite. It was obvious that, even in 1817, the colony was not fulfilling the expectations of many settlers. Their disillusionment was not ameliorated to any significant extent in the succeeding twenty years either, if the armed uprisings of 1837 and 1838 can be taken as indicators.

The early legislation of Upper Canada reveals the priorities of the colony's administrators. The appointed Legislative Council was the locus of ultimate power in the early days of the province (although the King in Council, in England, had a veto over legislation).[6] The Legislative Council set out to create the necessary instruments to ensure the growth of a loyal British colony only a decade after hostilities with the rebellious American colonies had come to a close. Although it would be fifteen years before there would be much demand from the mother country for primary resources,[7] the legislation passed by the council in the period 1792 to 1803 was heavily devoted to providing the administrative infrastructure for a successful colony. In the first dozen years of Upper Canada, the Legislative Council passed 108 acts, which can be categorized as in table 4.1.

It is significant that there was no allowance for a poor law, apparently because of Simcoe's dislike of the foundering arrangement for poor relief in England. In fact there was little effective legislation for the poor until after midcentury, although an early act allowed the indigent to be housed temporarily in district jails.[8]

Robert Gourlay recognized that the province was replete with settlers whose farming ambitions were unsatisfied: recall his depiction of frontier farmers "imprisoned in the woods," and, "after five or six years perseverance and hope, ... chilled with disappointment."[9] Gourlay ran into trouble with his thirty-one queries, designed to take the pulse of

Table 4.1
Legislation of Upper Canada, 1792–1803

Legal system	32
Regulating commerce	17
Assessments, taxes	4
Supplies	3
Property instruments	10
Preparation for assembly	4
Security	7
Municipal government (including roads)	9
Other	22
Total	108

Source: Wicksteed, *Provincial Statutes*

the province in 1817. He soon identified the source of the trouble as a comfortable, well-rooted oligarchy, among whom was the heartily disliked Bishop Strachan, Gourlay's "monstrous little fool of a parson."[10] Whatever the reason, the returns to Gourlay were far from complete. The pattern of the returns was regional, perhaps reflecting the variable influence of the York (Toronto) clique in the province. From the head of Lake Ontario to the west, there were heavy returns; the centre and the far east of the province sent almost no returns.

Gourlay's inquiry was largely concerned with the farming enterprise, but also with the conditions and facilities of rural society. In an attempt to give more substance and general applicability to the results of his survey, Gourlay provided a summation of the returns from the western third of the province. These returns accounted for all but half a dozen of the occupied townships between the Niagara and St Clair Rivers. He claimed the returns represented a population of 26,077 (2,000 higher than his tables; perhaps including military personnel) supported by the following amenities: 20 places of worship, 35 resident preachers, 20 medical practitioners, 132 schools, 114 taverns, 130 stores, 79 grist mills, and 116 saw mills.[11]

Although these aspects of the social fabric were not evenly distributed throughout this region, and although there is no way to test the accuracy of the data thoroughly, perhaps we can get an idea of the conditions at that time by some simple calculations. For example, there was an average of one school and one store for just over 200 people, and one tavern for every 237 people. There was one medical practitioner for 1,349 people, one grist mill for 341, and one saw mill for 233.

These calculations illuminate the conditions in the mixture of old and new areas in the western third of the province.

At the east end of Lake Ontario was a series of long-settled townships that also reported to Gourlay. They accounted for a large part of the original Loyalist heartland, although one – Wolford – was a more recent settlement in the second tier of townships back from the St Lawrence River. The other four townships in the series – Ernestown, Thurlow, Lansdowne, and Charlottenburg – were on the shore of either Lake Ontario or the upper St Lawrence. The calculations seem to illuminate the generally older, more developed character of the settlement. There was an average of one school for 134 residents, one store for 123, and one tavern for 95. There was one medical practitioner for every 584 people, one grist mill for every 292, and one saw mill for every 156.[12]

Identifying Deficiencies

Gourlay's dream of Upper Canada as a refuge for the poor people of Britain came to pass, despite the efforts to subvert his inquiry. Gourlay himself was exiled to the United Kingdom, but tracing his path in the opposite direction was a flow of immigrants that began to number in the tens of thousands by the late 1820s. These were people who came out of the social and economic turmoil caused by the extensive readjustments following the Napoleonic Wars. Most of them would have heard of the Poor Law, the debates over qualifications for the vote, the loosening of restrictions on organizing labour, and the rudimentary improvements in working conditions wrought by the periodic Factory Acts. They would also have been familiar with the anger aroused by the galloping mechanization of industry that put many people out of traditional occupations. The Old World climate of opinion on social change must have seemed jarringly alien in the superficially bucolic environment of Ontario before the mid-1840s. Some of these immigrants, in the spirit of Robert Gourlay, became gadflies of the colonial administration and reactionary elite, and they became spokesmen for change.[13] Even the mass of immigrants who had no partiality to radical action must have noticed the stultified society and entrenched privilege of Ontario. The disjointed uprisings of 1837 and 1838 were objections to a system seen to be unjust and out of step with its population.

As late as 1840 some facilities that might be thought essential, such as schools, were not available to a large part of Ontario's rural population. Governor General Lord Sydenham sent out a questionnaire to officials in the districts, – specifically, to sheriffs and to presidents of agricultural societies. The answers to the two sets of questions in each questionnaire were to be used for the enlightenment, on one hand, of "Emigrants with Capital," and, on the other, of "Emigrants of the Labouring Classes."[14] Regarding the emigrants with capital, question twenty-six read "Are there places of education for the children of the middle classes?" The chief emigrant agent for Upper Canada, in Toronto, replied: "the towns are tolerably well supplied, but schools in the rural districts, more particularly in the newly settled townships, are few in number, and of a very inferior description [*sic*] – but the subject of education now occupies the attention of the Government." There was a variety of opinion in the different districts, however. From Wellington District the answer was "in most of the tolerably well settled townships"; from Northumberland, "yes, in nearly all the principal towns." A full response from Sheriff Treadwell, in Ottawa District, pointed out that there was a government-endowed school in each district town as well as "a few excellent private schools," and the beginnings of universities in Kingston and Cobourg.

The answers to questions regarding "education in the rural districts" for emigrants of the labouring classes ranged from a rather vague "I believe there is in the whole of them" (Talbot District) to "there are very limited means in most of the back settlements" (Northumberland District). The chief emigrant agent enlarged on his earlier comment: "schoolmasters are much wanted in the rural districts – the schools are few in number."

Other facilities common in the old country were either rare or nonexistent in Upper Canada. Question thirty-five, for the information of the labouring classes, asked, "are there any hospitals or infirmaries?" The chief emigrant agent could answer briefly, "yes – there are Hospitals at Quebec, Montreal, Kingston and Toronto." Respondents in more remote areas did not seem to know. The sheriff of Talbot District had no answer, while Thomas Saunders in Wellington District said "No," and the respondent in Simcoe District said simply "Yes." Perhaps the answer in the minds of all three was "in practical terms, no," because Toronto was a difficult journey of sixty or more miles for most of the people for whom the respondents were answering.

In answer to the question "are there any Benefit Societies?" Sheriff Treadwell in the Ottawa District provided a characteristically full reply: "there are national societies ... in the principal Towns of Upper Canada. The St George, St Andrews, St Patricks and also a German Society ... These societies do much in a quiet and unostentatious manner to relieve the unfortunate and to extend the friendly hand to the Emigrant." Apart from the chief emigrant agent in Toronto, who also cites the national societies, most other respondents had no answer to the question.

Question thirty-seven asked, "is there any fund for relief of the destitute?" Most respondents from the rural districts said "no" or gave no answer. The chief emigrant agent enlarged somewhat: "Town Wardens are appointed to afford relief in cases of necessity – but relief is always confined to the helpless." Treadwell was of the opinion that "there are Charitable associations in the Chief Cities and Towns sustained by voluntary subscriptions. The provincial Government and the Inhabitants contribute liberally towards the supporting of the sick and the forwarding Destitute Emigrants free of Expense to the Rural Districts."[15]

A question on the existence of savings banks was answered variably, quite often with "No." The sheriff of Bathurst District replied, "none except in large cities," while the chief emigrant agent specified that there was such a bank in Toronto. The general lack of this kind of facility, commonly used by small depositors in Britain, led to the arrangements made by the Canada Company to help its customers save and transfer money to Upper Canada in the 1830s.[16]

The picture sketched by the reports from the various districts showed Ontario in 1840 to be generally in a rudimentary social condition. Outside the towns and the long-settled shorelines, there was very little in the way of a social infrastructure that would have been familiar to the numerous immigrants. It could even be argued that the outbursts of frustration in 1837 and 1838 were at least partly fuelled by the impatience of people who had waited overly long for this Promised Land to live up to their expectations. They knew that it was not unreasonable to expect a government to put effective pressure on absentee landowners, to extend good roads into remote areas where pioneers were encouraged to settle, and perhaps to erect school buildings where the majority of the population was found – outside the towns.

A year or so after the rebellions occurred, the British government dispatched Lord Durham, dubbed Radical Jack in England, to assess the situation. There were many in Upper Canada who found Durham's procedure and his report to be a concrete example of his nickname, and thus totally unacceptable. Even those more sympathetic, such as Sheriff Henry Ruttan, at least thought the report ill-conceived: "now that this Report of the High Commissioner [Lord Durham] has communicated its combustible materials to the flame – unless a caution and wisdom almost superhuman interpose, I much fear that a long period will elapse before perfect harmony be again restored."[17]

In 1840, Ontario was still coming to grips with the conditions that had given rise to the armed insurrections and with the "cure" that had been prescribed – Lord Durham's controversial report of 1839, leading to the forced union of Upper and Lower Canada. Notwithstanding Henry Ruttan's pessimism, it was only the following decade, with increased exposure of the colony (now Canadas West and East) to external scrutiny, that proved to be a time of real revolutionary change. But this decade was eventful indeed; apart from the initial entry of farm settlers into the pristine woodland, the 1840s were the most momentous pre-railway years in terms of measurable geographic impact.

At the opening of the 1840s, Ontario was dominated by tree cover and seasonal rhythms, and most residents laboured under some degree of isolation. By the end of the 1840s, the last of the truly farmable sections south of the Canadian Shield were being opened for settlement, and the province was being knitted together by an improved road network at the zenith of the age of the stage coach. In contrast to many earlier unfavourable comparisons with the neighbouring states, the secretary of the Board of Registration and Statistics could go on at great length to show how well the Canadas, especially Canada West, compared with the "go-ahead" Ohio Country.[18] The early population concentrations, especially those surrounding the three old gateways into the province, were being eclipsed by the magnetic growth of the centre – meaningfully named the Home District. The major development in the province was taking place along the northwestern shore of Lake Ontario. Regional differences, however, were not disappearing; rather, they were taking on more permanent complexions based primarily on the fundamental natural characteristics underlying the practice of agriculture, notably the length of the growing season and the soil quality. A settlement fabric was taking shape; the lineaments of a social geography were becoming discernible.

The colony caught the attention of London not only because of the rebellions in the late 1830s, but also because of the inclination of the imperial administration to loosen colonial ties and phase out colonial trading preferences. The home country remembered that the loss of the American colonies in the 1770s had not, despite expectations, meant loss of trade. The channels of business were too important to both sides.[19] The imperial system had long been an economic haven within which the colonies could profitably dispose of their raw materials. Emergence onto the international stage quickened the maturing of Canada West, but the transition to free trade brought a sense of vulnerability that diminished only with the signing of the Reciprocity Treaty with the United States in 1854.

One reality of entering the arena of international competition was the growing struggle between the St Lawrence Valley and New York City for the interior of the continent. Although serious proposals for railways in Ontario were made during the 1830s, nothing came to fruition until the 1850s. The union of Upper and Lower Canada in 1841 led to an emphasis on the canals of the upper St Lawrence River, so that by 1848 it was possible for vessels of proper ocean-going dimensions to enter the Great Lakes. A more auspicious connection with the outside world, however, was the single telegraph line that advanced from Buffalo along the northern side of Lake Ontario toward Montreal in 1846.[20] Suddenly there was a whole world to challenge and inform provincial society, and the newspapers in particular responded quickly to this new source of information. (See further discussion in chapter 6.)

Perhaps the rebellions and the profound reassessments that followed shook loose some of the deadening influences in the colony, encouraged rethinking in all realms, and precipitated the entry of new players (notably immigrants) onto the stage.[21] But access to new sources of information was the catalyst. Suzanne Zeller traces some of the scientific innovations that "elevated the Canadian imagination to new heights": the inauguration of the Geological Survey (1842) and the first Magnetic and Meteorological Observatory, in Toronto (1839),[22] combined with the popular communication of scientific information and Egerton Ryerson's attempt to include it in the evolving public school system. On other fronts, there was a gradual shift towards responsible government, and with it important initiatives such as the Common School Acts, the Guarantee Act (to make the financing of railway projects attractive to investors), and the first thorough Municipal Corporations ("Baldwin") Act during the second half of the decade.

There was a progression, that seemed to culminate with the dawning of the railway era, from an Upper Canada immersed in the crude struggle to tame a territory for farming to a society passionately engaged in "putting its house in order" – from parochial concerns for survival to provincial concerns for commonweal. Because the English Poor Law had been abjured with the founding of Upper Canada, it was necessary to construct a welfare system from scratch. An allowance for houses of refuge (or industry) for the indigent and poverty-stricken first appeared in provincial legislation in 1837, and although of negligible impact, this act was a signal of the stirring public conscience. The acceptance of some public responsibility for the treatment of mental disorders was manifest in 1841 with the establishment of the first dedicated facility (the Lunatic Asylum of Upper Canada) in temporary quarters in Toronto. A permanent building was begun four years later. Conveniences for immigrants had been maintained for some years in the larger centres, but these and many other of the embryonic institutions were overwhelmed by the huge, unparalleled influxes of Irish people at the end of the decade.

In addition to the government agencies to control and to succour, there was the growth of organizations based upon cooperation and upon common ideology or livelihood. Agricultural societies had begun to multiply in the 1830s, with government grants, and a provincial prize fair was first held in 1846. Self-improvement was encouraged through the mechanics' institutes and temperance societies that began to proliferate during the 1830s. Fraternal organizations of various stripes also became widespread by the 1840s.

The railway is often credited with being a kind of mythical prime mover that set in motion a chain of important events. After the middle of the century the railway did begin to transform the landscape, but the socio-political ground had already been well prepared. Profound changes that affected the way the society organized itself had been introduced and assimilated. The main body of this chapter focuses on a series of pictures of a society shifting from being based upon primary production (agriculture and timber) towards the beginnings of an industrial base. It was a process of growth for the society of Canada West, with both successes and failures. Occupance of an area leads to provision for what the society sees as essential functions. The next few pages elaborate some of the more notable and critical functions that evolved in early Ontario. A society seldom works harmoniously at all times, and Ontario was no exception. Some prominent and characteristic dysfunctions are also identified and discussed. An assessment of how well the territory was provided with social facilities is attempted by a measure of "social intensity" (figure 4.10). Social intensity, of course, is also a general measure of how much ecological change landscape recreation wrought in the various parts of the province.

LANDSCAPE AS SOCIETY'S RE-CREATION

A variety of initiatives began to be woven into the social fabric of the province. For most of peninsular Ontario the initial clearing of enough land for survival had given way by the 1840s to the beginnings of a coherent system for marketing farm surpluses.[23] The vignettes discussed in this section should be seen as samples of a broad-based evolution of settlement, and particularly of the effect of a society on the land through the creation of a *built environment*, which includes the preparation of land for growing farm crops as well as the construction of houses, barns, or factories. Because it was evolving everywhere, settlement embraced variations from a high level of development adjacent to the surveyed lines of the proposed railways to primitive, forested frontier fringes.

A Postal System

The evolving social structure was expressed on the ground by a spectrum of roads and trails. Closely related to the road network was the system of post offices. Access by mail was almost as important to early settlers as access by road. As shown in chapter 1, a post office had become a standard indicator of the "civilizing influence" of society at work in an area of new settlement. Postal service existed under the aegis of the government, and although the farther west the mail went, the slower and more unpredictable the service became, the system was being vigorously expanded in the early 1830s.[24] The existence of a post office and mail delivery can serve as a surrogate for a number of other conditions. Postal outlets were expected to be financially self-sustaining; thus, their survival says something about settlement activity roundabout, such as the density of population and the state of the roads. Because this study views a post office as part of the social landscape, the distribution of all the postal outlets provides one of the foundations for the display of this landscape in figure 4.10.

By the mid-1840s in Ontario there were 254 post offices, as reported by the deputy postmaster general, but – indicating the pace of change – there were about 500 post offices, reaching to the frontier fringes of the province, in 1850. This expansion had already begun nearly a decade earlier: of the forty-three post offices established in Upper and Lower Canada and part of New Brunswick between July 1837 and July 1841, thirty-five were in Upper Canada, a testimony to the attraction of the new areas despite the agitation that led to the Rebellion of 1837.[25] A map of the post office openings from 1837–41 (figure 4.1A) sketches out the areas where the main growth was occurring; that is, not at the frontier proper but directly behind it where the infilling brought a rapid increase in population and in cleared acreage. The base from which this expansion sprang is portrayed in figure 4.1B, the previous establishment of post offices.

Before 1828, post offices primarily traced the main corridors paralleling the St Lawrence River and Lake Ontario into the Niagara Peninsula, the Governor's Road west from the head of Lake Ontario, the less concentrated line of settlement along the Lake Erie shore, and in the east the cluster of early Scottish settlements. This is largely a consolidation of John Graves Simcoe's military plan and the founding settlements that benefitted from it. The expansion of post offices from 1828 to 1836 was highlighted by infilling and a few extensions away from the St Lawrence and Lakes Ontario and Erie; for example, the line into the Canada Company's Huron Tract in the west, the strong extension of

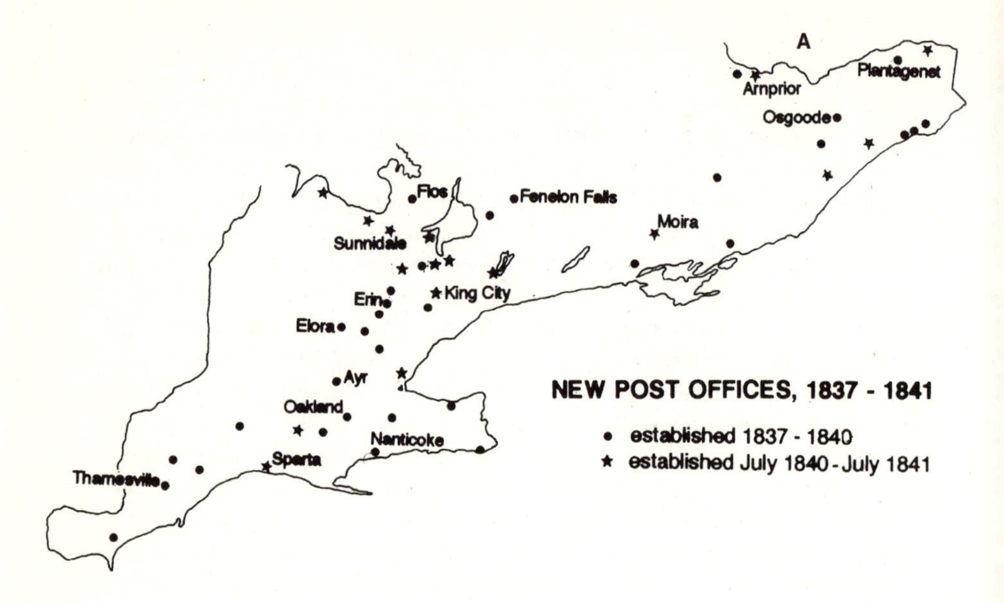

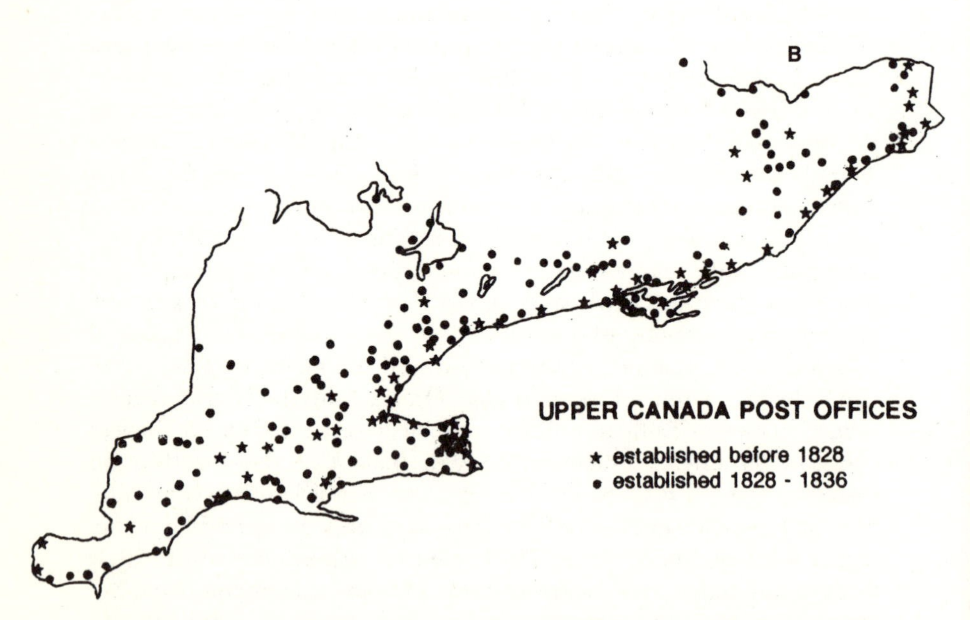

Figure 4.1 New post offices opened. (A) 1837–41: evidence of a moving frontier; (B) before 1837: a materializing social facility (Canada, *Journals of the Legislative Assembly*, 1846, appendix F, nos. 16, 17, 19)

the Yonge Street corridor beyond Lake Simcoe, the string of villages along the Trent River system north of Rice Lake, and the push up the Ottawa Valley. This development set the scene for the final expansion onto agricultural land south of the Canadian Shield, although persistent low-development areas in the east, on the southern salient of the Shield near Kingston, and in the far southwest would not be wiped out by mid-century (figures 4.10 and 8.1 show relevant later conditions).

Functions

In a colony predominantly engaged in farming or related pursuits, it is not surprising that the earliest organizations, apart from religious ones, were agricultural. Although many early initiatives – whether commercial or social – tended to be ephemeral, agricultural societies gained a large measure of permanence in the 1830s through legislation that provided financial support. A number of county societies were established by the 1830s, but the 1840s saw an outburst of local societies, even at the township level (figure 4.2). Generally the agricultural societies established before 1840 were located adjacent to the principal urban centres; those set up during the 1840s extended well into the hinterland, including some recently settled townships within sight of the Canadian Shield in the centre of the province, as well as in the Huron Tract on the road to Lake Huron. The societies formed as a result of the 1830s legislation usually held annual or semi-annual prize fairs aimed at improving agricultural products. The act that sanctioned annual grants to agricultural societies outlined the responsibilities of a society: "importing valuable livestock, grain, grass seeds, useful implements, or whatever else might conduce the improvement of agriculture."[26] Breeding animals would be rotated among the farms of members and seeds would be distributed for trials by members.

The societies' devotion to agricultural improvement indirectly contributed to social betterment. The fairs sponsored by the agricultural societies were different from the markets that had sprung up in formal and informal venues across the province. The exchanging or selling of goods, being one of the most primitive of human interactions, was carried on from the first days of settlement. The sale of farm products began to be institutionalized by 1800; market locations appeared on the ground plans of many Ontario towns, and market regulations were passed by municipal governments.[27] The agricultural fairs of the provincially funded societies went a step beyond basic commerce, although the stock and produce were often available for sale after the judging and allocating of prizes, and sometimes the fair was combined with a one- or two-day market. Frequently the distinction between the

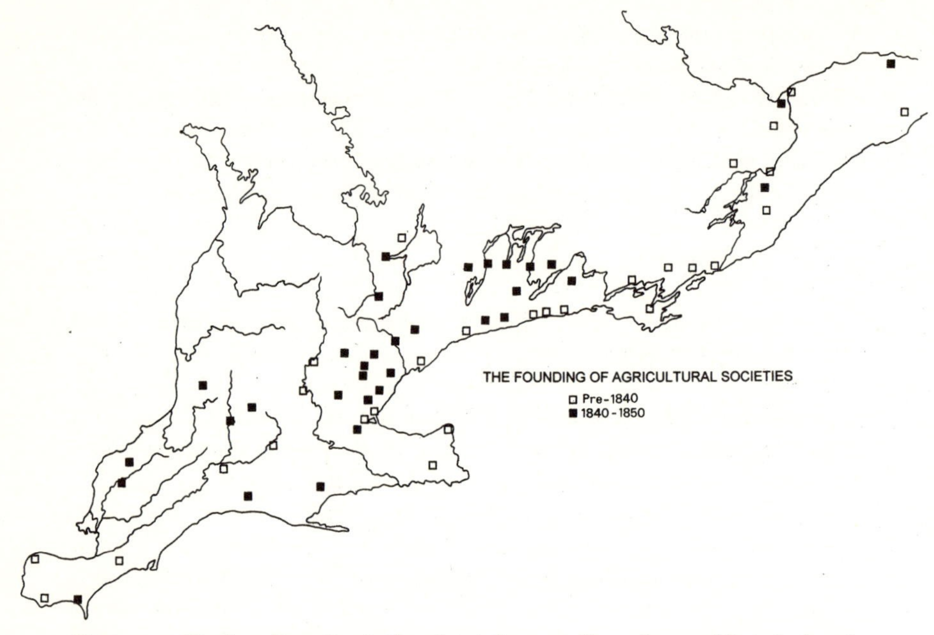

Figure 4.2 The founding of agricultural societies: pre-1840; 1840–50 (Ontario Association of Agricultural Societies, *Story*; Talman, "Agricultural Societies," 545–52)

different kinds of fairs and markets was blurred.[28] The improvement in the quality of products, it was sometimes argued in petitions for market fairs, could also be brought about by "comparison shopping" at the fairgrounds.

Another medium for improvement, comparable to the agricultural societies but meant primarily for urban workers, was the mechanics' institute. The intention behind this organization was to provide a kind of rudimentary polytechnic, or Workers' Educational Association, with the expectation that the improvement in craft skills would be disseminated. In addition, the most influential proponents saw the institute as a means of upgrading the morals of the mechanics. By diverting the working man from intoxication and dirty habits it would "elevate, ennoble and enlarge the mind."[29]

The first mechanics' institute in Upper Canada was established at York in 1830. It was followed by another six by 1839 in towns from Kingston to Niagara and London. Founding accelerated in the 1840s, with seventeen institutes inaugurated in a widening territory from Bytown (later Ottawa) to Stratford, and Mitchell, near Lake Huron (figure 4.3). The founding of mechanics' institutes reflected a "spirit of improvement" moving through the province in the 1840s, particularly

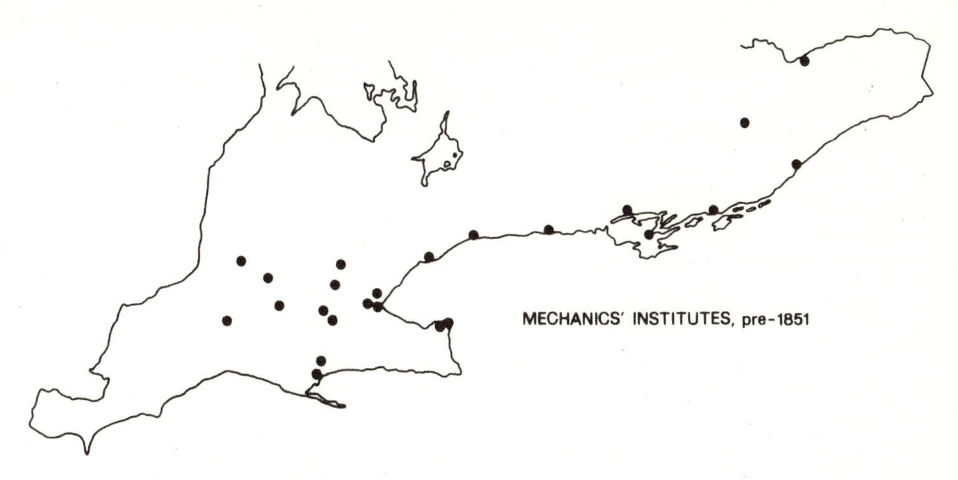

Figure 4.3 Mechanics' institutes founded 1830–50 (Vernon, "Development of Adult Education")

during the economic recovery after 1842.[30] This was the same stimulus that led to the outburst of railway building, and the legislation that expedited it, a few years later.

The proliferation of mechanics' institutes was evidence of movement towards social structuring. That the institutes never effectively captured the interest of the artisans and workers whom they were supposed to "improve" provides a lesson in the fruitlessness of social structuring superimposed by one class upon another. Either the activities of the institutes were not attractive to most "blue collar" workers, or the paternalistic tenor of the organizations repelled eligible members. Lectures were offered on various uplifting topics in the early years of the institutes, but in most cases it was the lending library of the institute that became its major function. The "improvers" became dissatisfied with the role of the mechanics' institutes because the books most in demand were fiction or other kinds of "lightweight" literature. The province transmuted all mechanics' institutes into public libraries in the 1890s; hence the institutes played a role in the social geography of Ontario, but not the one originally intended.[31]

More "grassroots" in nature than the mechanics' institutes were the fraternal societies and clubs that began to appear before the middle of the nineteenth century. The rapid spread of the Orange Order after its formal inauguration in Brockville in 1830[32] suggests that this fraternal organization met a need for assembly and common activities in most parts of rural and small-town Ontario. There was no other fraternal organization at this time in Ontario that came close to having the success of the Orange Order. Whatever our judgment of its purpose and its ultimate influence, it was probably the organization most representative of the population of mid-nineteenth century Ontario, being predominantly monarchist, Protestant, and conservative. And to the benefit of historical scholars, the Orange Order left records that allow a reconstruction of its activities.

The spread of Orange Lodges in the 1830s and 1840s can only be described as mercurial (figure 4.4). Within two decades lodges had been established from the Ottawa River to Lake Huron, and from Lake Erie to the edge of the Shield. The only sizable gaps in the pattern of lodges in the well-settled parts of the province were the areas of heavy Roman Catholic settlement, such as the extreme southwest (mainly French) and the extreme east (French, Scottish, and Irish); there were smaller gaps in the Grand River Valley and south of Lake Simcoe.

The Orange Lodge was an organization that could arrange a variety of relevant activities for a farm population beyond the stage of frontier survival. Cecil Houston and William J. Smyth point out the social importance of the celebrations (especially the July parade), the concern and assistance for "distressed members," the attendance at funerals of members, and the bonhomie of the monthly meetings – all framed by a common adherence to a somewhat bigoted Protestant view of the world.[33] There would also have been many informal relationships arising from membership.

By the 1840s, access to medical treatment began to be one of the measures of the quality of life. At the time of Gourlay's survey in 1817, medical practitioners were irregularly distributed across the province – from six in the small district of Niagara and three in the expanding Gore District to six in the London District and five in the Western District, for example. By the time of William Smith's survey in 1850, there was an abundance of practitioners in most areas (figure 4.5).[34] The apparent lack of doctors in the eastern and western extremities of the province reflects lower densities of settlers because of expanses of ill-drained or infertile land. The thinning of practitioners towards the settlement fringes is not surprising, but it is worth noting that doctors were available in most of the frontier "jumping-off" points for the Queen's Bush and the southern Shield, such as Goderich, Saugeen (Port Elgin), Sydenham (later Owen Sound), Orillia, Beaverton, Peterborough, Tamworth, and Perth.

Medical doctors formed an identifiable group, for a system of licensing had been in place from the beginnings of Upper Canada. An act of 1827, which remained in force into the second half of the century, attempted to maintain standards of medical practice comparable to those in London and the British military through a medical board that

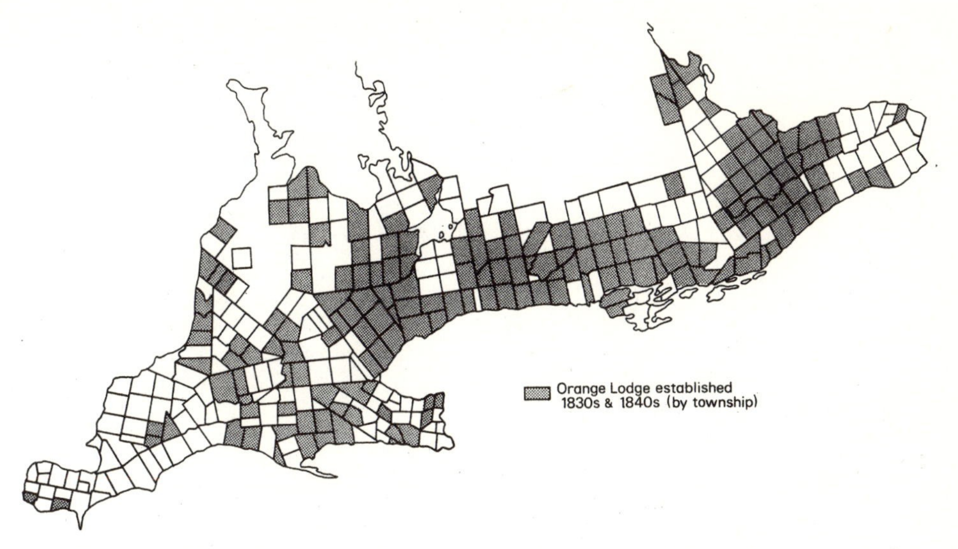

Figure 4.4 Orange Lodges established, by township, in 1830s and 1840s (Houston and Smyth, "Orange Order")

examined applicants.[35] Not all physicians were rooted in one place, however; like the settlers they treated, individual doctors could be transient. One example is Dr J. Forbes, who in 1850 was in St Catharines, but in 1851 was a doctor in Chippawa. Also, in the year or so between William Smith's and Robert Mackay's listings of professional men, a Dr Henry Goodman had moved from Cayuga on the Grand River to St Catharines. In that same period the number of doctors in Stratford, at the eastern entrance to the Huron Tract, increased from one to three.[36]

A well-known development in Ontario is the rise of a school system in the 1840s to replace what had been a multifarious hodgepodge of local arrangements.[37] Whether or not it reflected "maturation" of the society, educational reform was part of a broad movement in the Western world towards the bureaucratization of society and the evolution of distinct professions. The major proponent of the Ontario school system was Egerton Ryerson, who had personally investigated a variety of systems in the British Isles, continental Europe, and the United States.

The Common School Acts of 1841 and 1843 promoted the power of a central educational administration through the establishment of a bureau with increasing funding and the means of collecting and coordinating information. But it was with the act of 1846 and the appointment of Ryerson as chief superintendent for Canada West that centralization and standardization began in earnest. Ironically, as the system became

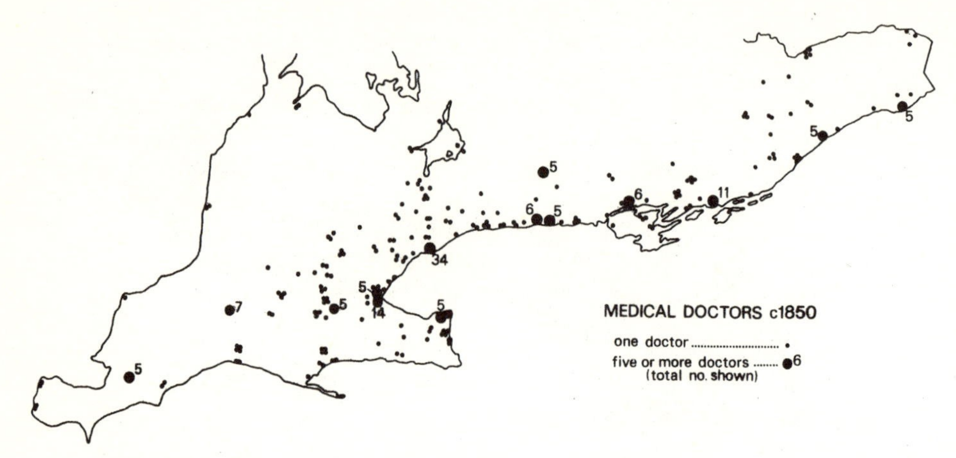

Figure 4.5 Medical doctors, ca 1850 (Smith, *Canada*, 1,2: business directories)

more public – that is, governed from the capital in terms of administration and curriculum, and with a hierarchy that reached down to the township level – it became less popular, as it lost local flavour and variability. Centralization provided the authority and financing for schools which, in some cases, would not otherwise have existed.[38]

During the transition from the previous mixture of government-aided and -unaided schools to the provincewide, publicly funded system, a good deal of reorganization took place. New schools were opened, and others (remnants of the previous arrangement in which perhaps a majority did not receive government grants) were closed or amalgamated. Some districts had fewer schools in 1849 than in 1847, and others had about the same number. Not surprisingly, districts that had been recently opened or whose fringes were being settled experienced an increase in the number of schools. Spectacular increases were recorded in the new districts: from 41 schools in 1847 to 65 in 1849 in Huron; from 83 to 126 in Simcoe; and from 73 to 98 in Colborne.[39]

The extension of the public school system into all parts of the province during the 1840s was probably the most thorough-going expression (after the postal system) of the extension of an effective social structure – indeed, a modern-style bureaucracy – over the whole of the territory. The school system touched virtually everyone in a way that the earlier institutions, such as the courts, churches, banks, or even assembly representation, had not done.

A good indication of the robust functioning of a society is the flourishing of creativity and innovation. Patents for inventions began to be registered in the Canadas in 1824, but none was registered for Upper

Canada until 1831.[40] Over the next nine years the increase in entries was very gradual, but by the 1840s there were a dozen or more patents a year registered in Canada West. The distribution of innovation (figure 4.6) reflects the density of population, as one might expect. But there is a marked contrast between the shore of Lake Ontario, where the development of industry was under way, and the shore of Lake Erie, where there was little indication of industry or innovation, echoing Kenneth L. Sokoloff's findings for New England: "the growth of manufacturing productivity (especially in less capital-intensive industries) and of patenting appear to have spread out together from urban districts after 1820, along with the extension of transportation networks and extensive involvement in inventive activity by individuals with rather ordinary skills and backgrounds."[41] In Ontario there was a flurry of innovation in the vicinity of some of the main urban centres, and an impressive amount of creativity also emanated from the intermediate rural belt between the Great Lakes littoral and the fringe of settlement. Potential inventors on the frontier and in some of the more difficult areas for settlement in the west and the east seem to have been preoccupied with the challenge of survival, although one might also suspect the inadequacies of the means of communicating.

The patents have been distilled into four general types. "Household, quality of life" includes such things as improvements in cookstoves, heating and cooling equipment, the cleaning of clothes, medical and therapeutic apparatuses, and musical instruments. "Farming equipment" is self-explanatory, but during this period inventions were focused on machines for threshing, cleaning, and grinding grain, as well as churns and tanning machinery. "Power machines, mills, heavy construction" embraces bridge construction, modifications of steam propulsion, lumber milling, pumps and hydraulic applications, and ship construction. "Other" includes improvements in saws, bricks, nuts and washers, guns, and equipment for fire extinguishing, distilling, and brewing. Among the urban centres, Toronto's contribution was varied, whereas Hamilton had some emphasis on household and quality of life inventions, and Kingston had a relatively large number in the "power machines" category. The rural belt contributed a variety of innovations, although there was an emphasis on power machines and heavy machinery around the Grand, St Lawrence and Ottawa Rivers.

Dysfunctions

Even if one considers only the Rebellion of 1837, clearly, all was not well in Upper Canada. It is significant that, at a time when profound

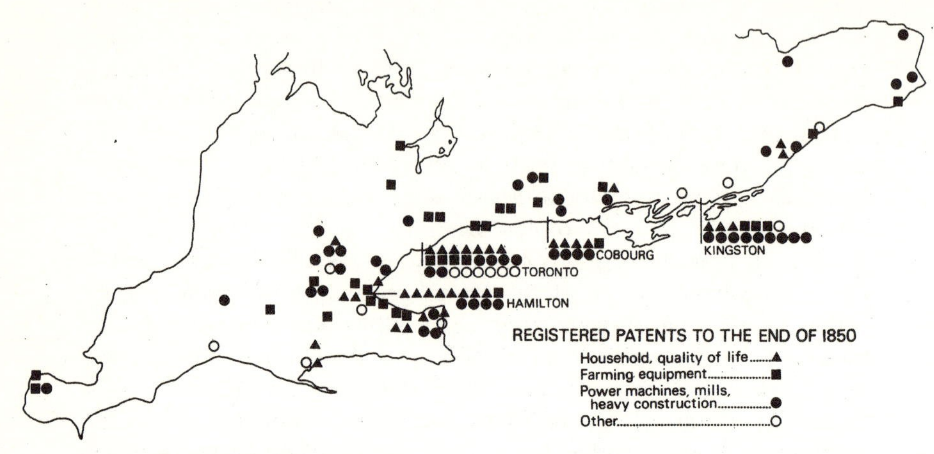

Figure 4.6 Distribution of innovation: residences of persons registering patents, 1831–50 (Canada, Patent Office, *Patents*, vols. 1, 2)

changes were rocking society in the old country, it appeared to many people in Upper Canada that change would never overtake the old guard and the merchants who dominated the colony. The outcome of the rebellion was inconclusive, but the wide distribution of disaffection was clear; arrests on charges related to rebellion occurred in almost all of the districts from Western to Midland (figure 4.7). What is of interest to the present study is that most of the resentment of the status quo seemed to be expressed in those parts of the province that had been advancing most vigorously in standard of living – primarily the belt of townships north of the first tier of townships but south of the frontier areas. This was territory, settled mainly after 1820, where farming had begun to prosper. A society "on the make" can be impatient in the face of a stultifying administration and a privileged class, even to the extent of violence. On the other hand, such a society can also harbour "casualties" who have nothing to do with armed uprising; these are the people who represent the social dysfunctions.

There seems to have been some resistance to setting up public facilities for relieving poverty, perhaps because it was felt that this problem was being imported along with destitute immigrants, against the wishes of the colonial population. The ubiquity of poverty in British North America in the 1840s has been documented.[42] The extent to which Upper Canada had been prepared to go was to allow the destitute to be lodged in the district jail, but this was normally a short-term arrangement because the poor were generally thought to be largely responsible for their condition. As with most of the dysfunctions, there was a strong public prejudice that what was needed was not just shelter

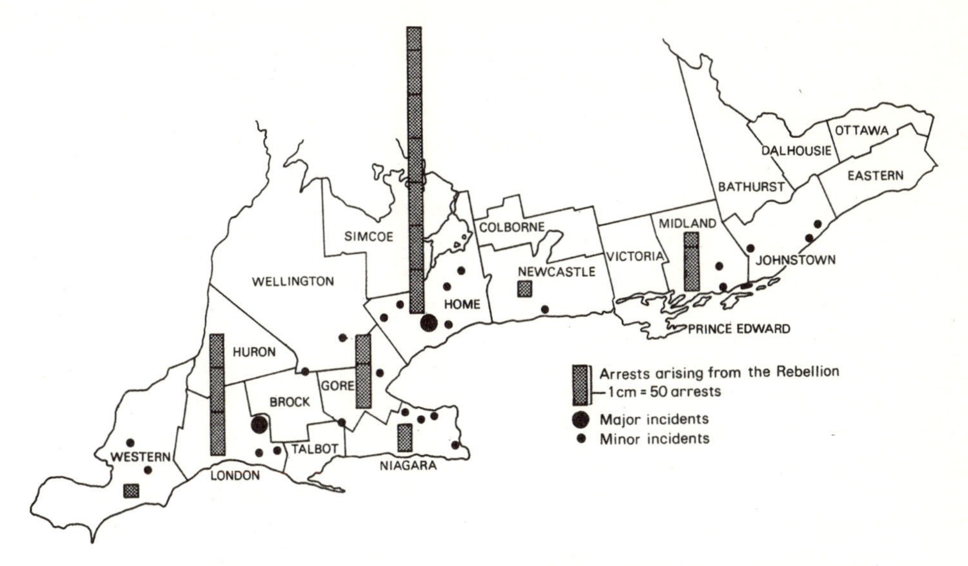

Figure 4.7 Rebellion as social dysfunction: incidents of a "rebellious character" (Arnold, "Geography of Colonial Unrest"). Minor incidents include assistance to rebels, attacks on postal carriers, reform petitions, casting bullets.

but also the correction of moral or psychological faults.[43] This prejudice lies behind the fumbling over a name for the institution to deal with the destitute: a House of Refuge or a House of Industry? The House of Industry Act was passed in the legislative assembly in 1837, but its terms were so unwieldy that nothing resulted. Charity had been provided over the years through voluntary donations, and various ad hoc organizations had operated, especially in the towns. Kingston and Prescott, both of which dealt with waves of immigrants, many of whom fell on hard times, gave charity to the indigent in the 1830s.[44] In terms of social geography in 1840, however, there appears to have been only one example of a house of industry – a privately funded institution established in Toronto in 1837. The House of Industry Act of the same year, although ineffective, was an exhortation that led to a rapid increase in examples of enabling legislation and action. These acts at least gave "a clear indication that by the mid-1840s the province was beginning to concede the necessity of public support for those in need."[45] Indeed, in the 1840s the government was financially assisting the Toronto House of Industry.

In Victorian Ontario, as elsewhere, employment did not necessarily remove the spectre of poverty. In fact, it was not uncommon to be seasonally employed. This might be expected among farm labourers and woods workers but, as revealed by the chief emigrant agent for Upper

Canada in 1840, there were many other vulnerable occupations. His table on wage rates gives an annual wage for nine occupations and daily remuneration for another nineteen, two of which are building trades explicitly described as "not employed during winter" (see figure 5.7 and related discussion in chapter 5).[46] At the top of this wage scale was the millwright. At the bottom of the scale – earning only from one-quarter to one-third as much as the millwright, and susceptible to seasonal layoffs – were farm labourers, tanners, quarrymen, carters, and dairy women. Other sources, however, claimed that a "day labourer" (perhaps on canal construction) could earn about as much as a millwright in a year. Whereas teachers ranged all the way from the lowest on the scale to higher than the millwright, the chief superintendent of education earned about ten times the income of a well-paid teacher.[47]

A closely related aspect of occupational vulnerability was illiteracy. It is very difficult to get a clear view even of the illiteracy that exists today, let alone that of the remote and shadowed past, but the historical record sometimes offers a glimpse. The members of militia units maintained in the aftermath of the rebellion had to sign a list individually to indicate the receipt of pay. From a sample of fifteen lists it appears that about 7 per cent could not sign their names. Militias around the Head of the Lake (the western end of Lake Ontario) showed an even higher percentage: four of seventeen members of the 3rd Gore Militia, stationed on the Burlington Canal in December of 1837, were unable to sign their names; and of thirty-seven members of the Dundas Militia extending their service in 1838, eight could not write their signatures.[48] These proportions are a minimal measure of illiteracy, of course, because being able to sign one's name is not a sure indication of being able to read and write. Nevertheless, one can visualize a general undercurrent of illiteracy in the population across the province, located largely in poorer, perpetually transient families. Although it has been argued that, even among successful citizens, illiteracy was not uncommon and apparently did not carry a stigma,[49] it is hard to accept that "the vast majority of Upper Canadians seem to have been satisfied with a minimal schooling." Perhaps the rapidity of the development of the educational system belies an apparent complacency.[50] Reassessments of elementary educational opportunities before 1840 have found that conditions were not as universally bad as the school promoters claimed. There was an effort made in most areas to provide access to at least basic schooling, and for some areas the bureaucratic system of the mid-century, with its distant, inflexible rule, was no improvement.[51]

Nevertheless, the level of "book learning" in Ontario was roundly criticized by many published writers on the eve of the establishment of the public system of education in the 1840s. Commonly the critics were visitors; Anna Brownell Jameson and Lord Durham are well-known examples, but similar opinions were voiced by a wide range of residents.[52] Most parents schooled in other countries recognized that their children were generally unable to get as full and as varied an education in Ontario as they themselves had received. What the critics were saying, it seems, was not that the rudimentary education available in the plethora of small schools in Ontario was insufficient for earning a livelihood as a farmer, but that this kind of education was not good enough for a society that was hoping to modernize and industrialize. The Common School Acts of the 1840s and later, and the efforts of individuals like Ryerson, while imposing on the province an increasingly centralized bureaucracy, gradually introduced an educational foundation sophisticated enough to sustain an industrial and commercial economy competing in international arenas. Whether it is considered a dysfunction or just a deficiency, the low level of educational provision was widely seen as a problem that needed to be addressed.

Critics were also voluble about another ubiquitous weakness in Ontario society – the overuse of alcohol. Susanna Moodie's description of the drunken denouement of a building bee is well known, and it was echoed by many other writers, including William Thomson, a participant in a bee in the early 1840s: "the neighbours came by twos and threes … and as they came up each got a drink of whisky out of a tin can … At first they went to work moderately and with quietness, but after the whisky had been handed about several times they got very uproarious – swearing, shouting, tumbling down, and sometimes like to fight … most of them were drunk and all of them excited."[53]

This does not fully establish the ubiquity of drunkenness, but another writer described "every inn, tavern, and beer-shop filled at all hours with drunken, brawling fellows, and the quantity of ardent spirits consumed by them will truly astonish you … the old settled farmers and Americans … drink like fishes all day, and for aught I know, all the night too."[54] In response to a question about the beverage of "the common labourers" in 1840, a local official in Woodstock replied that "Whisky is more common drink than Beer but the latter can be had at all times … for 30 s. per Barrell" (of 32 gallons), while another response from Medonte Township, northwest of Lake Simcoe, stated: "Beer not common enough. Whisky too cheap and in general use." An official in the thirteen-year-old settlement of Guelph reiterated that "whiskey is the common drink it can be had at 1/3 per Gallon." (This would make the price per barrel about 40 shillings.)[55]

Intemperance is as hard to measure as illiteracy, but it is possible to cite some enlightening circumstantial evidence. Figure 4.8 illustrates the distribution of distillers and brewers that appeared in William Smith's business directory at the end of the 1840s. One suspects that there were many other less "respectable" operators who were responsible for the manufacture of inferior products, such as a whisky that "smelled most horribly ... of the same quality as that known in Aberdeen by the name of 'Kill the carter,'" in the words of William Thomson.[56]

Distilleries were distributed from one end of the province to the other, roughly reflecting the density of population. There were more distilleries than breweries, with a striking concentration of the former in the middle Grand River Valley. Distillers were established near all of the expanding settlement frontiers, from the Lake Huron shore across the bottom of the soon-to-open northwestern corner of peninsular Canada West to the neighbourhood of the Shield edge east of Lake Simcoe and beyond to the Frontenac Axis. There was a puzzling lack of distillers in the eastern sector, with the only representatives appearing in the Scottish centre and at three locations near the St Lawrence River. One wonders if the fabled drinking prowess of the Ottawa watershed lumbermen was fuelled from these few outlets or from outside the region.

The apparent success of the distilling enterprise and the impact of its product were confronted from one end of the province to the other during the 1830s and 1840s by organized temperance. Following the example of the United States and Great Britain, Upper Canada formed the first temperance society in 1828. The initial thrust was to reduce the consumption of "ardent spirits" in favour of more wholesome beer. Later, the temperance movement, like education, became yoked to the ideology of progress, in which discipline, industry, cleanliness, and godliness were the acceptable personal attributes, and teetotalism was the admonition. The temperance movement in Ontario, according to F.L. Barron, began in spontaneous local organizations. The success of its mission, which spread rapidly in the first few decades, rested largely upon the strength of Methodism and the organizational resources of the towns and the areas of relatively dense settlement.[57]

Temperance arrived from the east. The major conventions and organizational inspiration were focused in Montreal during the 1840s. Figure 4.8 shows that temperance groups in eastern Ontario far outnumbered producers of alcoholic beverages. Is it possible that there were few distillers and brewers in that region because of the moral purgative of the temperance movement? The expansion of the Sons of Temperance, primarily a young men's organization, was also much stronger in the eastern part of the province. From its first

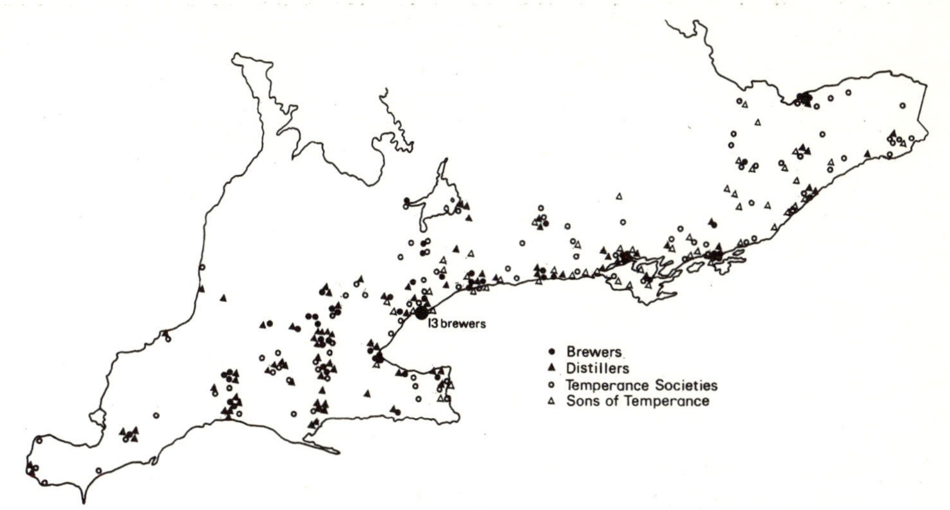

Figure 4.8 The standoff between alcohol and temperance: brewers and distillers, 1850; temperance societies active in 1844; Sons of Temperance branches founded 1849–50 (Smith, *Canada*, 1, 2: business directories; Barron, "Genesis of Temperance")

appearance in the province in 1848, the Sons of Temperance spread westward, reaching the head of Lake Ontario by 1850. The older temperance societies were found all across the province, in many cases apparently "checkmating" the distilleries and breweries, but in a few cases erecting advance posts ahead of the alcohol producers. In the backwoods, however, temperance did not have a significant presence, at least in terms of societies. Temperance was more characteristic of the advancing farm areas, where progress was the ethic, and modernization was the byword. The conservative elite did not welcome the wave of temperance advocates: Richard Bonnycastle complained in 1841 of "the meddling of these persons."[58]

The connection made by temperance advocates between intemperance and disorder could not be brushed off lightly, although it was probably more symptomatic than causal. Disorder appeared in its grossest form in the uprisings of 1837 and 1838, focused on certain parts of Upper Canada. Crime in general, however, was more randomly distributed. In the decade following the rebellion, criminal charges were made at an annual rate of around seven or eight per 10,000 people.[59] Individual districts could range well above or well below average as a result of untypical local conditions. The London District was above the mean in all three of the years assessed in table 4.2, while the Western District was well above in 1842 and 1846. On the other hand, the Home District, which usually recorded the largest or

Table 4.2
Criminal indictments in early Victorian Ontario
(by district and number per 10,000 population)

	1838 Assizes[a]			*1842 Assizes*[a]			*1846 Assizes*[a]		
District	*Indictments: Spring*	*Fall*	*Crime Rate (per 10,000)*	*Indictments: Spring*	*Fall*	*Crime Rate (per 10,000)*	*Indictments: Spring*	*Fall*	*Crime Rate (per 10,000)*
Western	4	7	5	23	7	12	32	9	15
London	12	15	11	16	12	9	50	13	16
Talbot	9	1	11	4	3	6	6	?	–
Brock	–	–	–	1	1	1	7	?	–
Huron	–	–	–	0	1	1	15	?	–
Niagara	10	18	9	11	36	13	19	6	6
Gore	14	22	7	17	12	6	18	18	6
Wellington	(in Gore)		–	2	1	2	7	?	–
Home/Simcoe	23	14	5	27	29	7	69	26	9
Newcastle	5	6	3	7	16	7	9	4	3
Colborne	(in Newcastle)		–	0	14	10	?	5	–
Prince Edward	4	3	5	3	1	3	?	8	–
Victoria	–	–	–	3	8	8	2	5	4
Midland	32	15	18	44	37	23	9	14	5
Johnstown	22	4	8	4	5	3	2	10	3
Eastern	0	2	<1	6	7	4	3	5	2
Ottawa	0	1	1	2	0	3	?	4	–
Dalhousie	–	–	–	–	25	[b]	4	16	9
Bathurst	5	0	4	5	10	7	1	10	4
	Total 248		Mean 6	Total 400		Mean 8	Total 409		Mean 7

Sources: Archives of Ontario, RG 22, ser. 134, vols 7, 8, 9; ser. 135, vol. 6. Population for 1838 and 1842 from Canada, *Census of 1870–71*, vol. 4; for 1846, extrapolated on semi-log graph from 1842 and 1848

[a] the assizes sat twice per year in each district town to pass judgment on criminal indictments brought by a grand jury. Crime rate rounded to nearest integer

[b] It is not apparent if this should be viewed as a full or a half year of cases. On either basis it is high; if a full year, the rate would be 15 per 10,000

second-largest number of indictments, was close to or below the mean rate for the province during these years. The Midland District had a surprisingly high rate in 1838 and 1842 for its mid-range population. The 1838 rate is accounted for by treason trials, but the story of 1842 involves multiple larcenies. The Niagara District's high rate in 1842 was largely a result of rioting in which most of those charged had Irish surnames. The eastern end of the province was generally pacific except for one hot spot at Bytown (later Ottawa). The Dalhousie District assizes were only activated for the fall of 1842, at which time twenty-one of the twenty-five charges dealt with rioting. The lumbermen's rivalries (sometimes between Irish and French Canadians, at other times Protestants and Catholics)[60] apparently were kept within the bounds of the district, because surrounding districts were markedly below the provincial mean. The fringes of settlement, such as Huron, Colborne, and Wellington, had records that varied from year to year, as did most of the districts. For the three assizes available, however, Wellington had no outbreaks of crime, whereas Huron showed an elevated rate in 1846 (figure 4.9).

The distribution of crime varied from time to time, as figure 4.9 illustrates. The Midland and Western Districts had crime rates that were often above the mean during the periods under analysis, whereas the eastern end of the province had rates that were at or below the average. The centre of the province was most often at the mean. Seasonal variations, however, cannot be investigated without finding details on individual cases, because holding cases over to later assizes was not uncommon. Also, delays in organizing cases, especially those in which a number of people were involved, muted any seasonal differences in the crime rate.

By far the most common crime was larceny, amounting to 347 of the 1,050 indictments in the three years. Unfortunately this term covered a wide range of theft, and apparently had not yet acquired the restricted modern sense of serious or "felonious" stealing. At the fall assizes in Niagara in 1842, one larceny conviction drew a penalty of four months in the district jail, while four others drew three years in the provincial penitentiary. There were some kinds of larceny that were singled out, such as stealing a horse or cow, stealing timber, or stealing a saddle. Stealing a horse (twenty-two indictments) commonly led to five years of hard labour in the penitentiary. Burglary (twenty-two indictments) sometimes drew the death penalty, but, failing that, a lengthy incarceration. Two men convicted of arson (out of a total of six indictments) in Talbot District, in 1846, were each committed to fourteen years in the penitentiary.

Canada mirrored the old country tradition of severe punishment for harming or threatening private property.[61] Damage to persons,

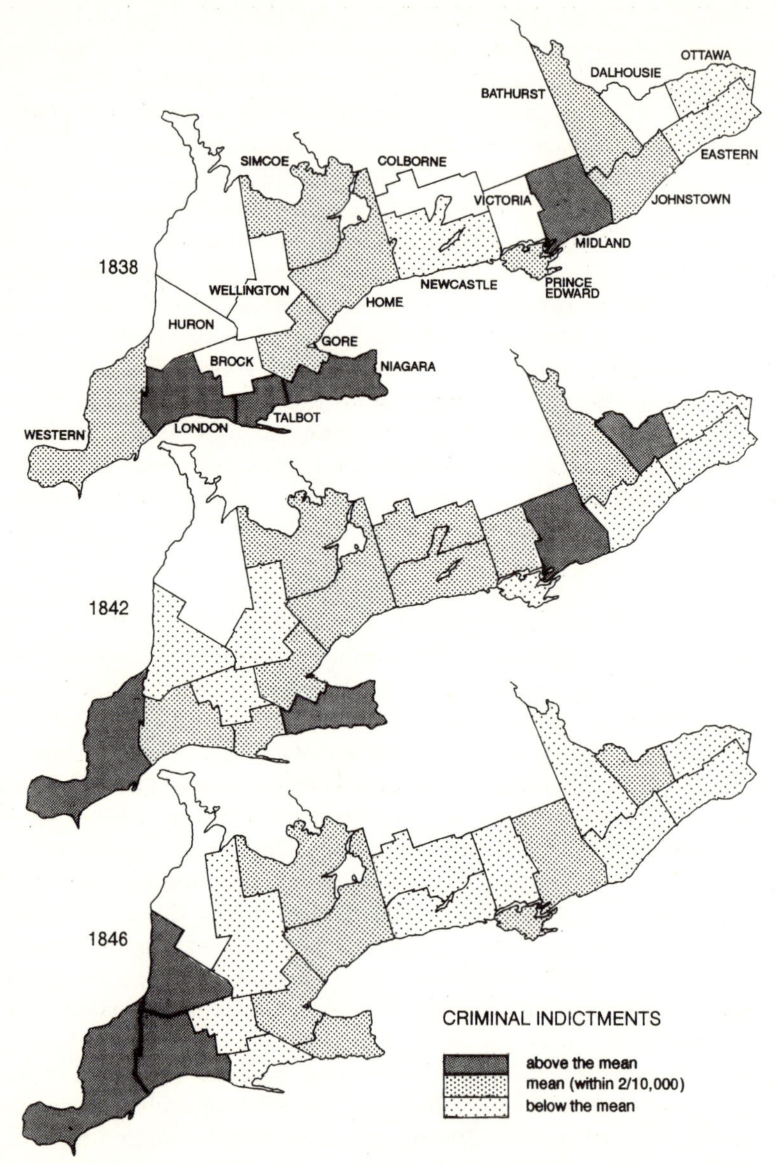

Figure 4.9 Criminal indictments by district, 1838, 1842, 1846 (cf. table 4.2) (Archives of Ontario, RG 22, ser. 134, vols 7, 8, 9; ser. 135, vol. 6)

however, was treated quite variably. Murder, manslaughter, and infanticide accounted for sixty-eight indictments. Sentences for murder ranged from less than a year in the district jail to the more common punishment of hanging. For a proven case of rape (thirty-five indictments for rape or attempted rape), the death penalty was usually meted out. One conviction for so-called "assault with intent to ravish," in Brock District in 1842, resulted in six months in the district jail; another in Simcoe District in 1846 led to eight months in jail; but a similar charge in Niagara in 1842 brought two years in the penitentiary.

The second and third most common indictments after larceny were misdemeanor and riot (including "riot and assault"), each of which accounted for a little more than one-quarter of the number of larcenies. Like larceny, misdemeanor was an unspecific term that occupied a segment of the legal landscape in contrast to felony, another unspecific term. A felony (seventy indictments) was a very serious crime that traditionally had drawn the severest punishments, and in Canada West in the 1840s it could mean execution. Misdemeanor, on the other hand, was considered less serious, and usually led to a few months in the district jail.

Riot, on the other hand, was a more clearly defined crime that appeared in connection with outbursts in different parts of the province in the years studied. In Johnstown, in 1838, there were fifteen indictments for rioting (perhaps treated as treason in neighbouring Midland District). In Niagara, in 1842, seventeen men were charged with rioting (periodic flare-ups occurred among workers on the canal) and in Dalhousie twenty-one were charged (probably from one of the episodes in the so-called Shiners' War).

Differences in the mix of indictments from place to place in the province may be in part a reflection of the different legal authorities in the districts, but they also hint at regional distinctions in society and the kinds of activities taking place. In the centre of the province, in the Home and Gore Districts, larceny was the overwhelming crime. With an average rate of criminal indictments, in which stealing was the common crime, Gore had an unusually large number of indictments for murder and manslaughter (even more than the more populous Home District). The crimes in London District were almost evenly split between larceny, misdemeanor, assault, and conspiracy. The Western District, while dominated by larcenies, also had the largest number of felony indictments (thirteen) of any district. The Midland District had a sample of almost every crime on record, but larcenies predominated and high treason and related lesser charges made a significant appearance in 1838.

Only eighty of the 1,050 indictments were handed down to women, but almost all the districts recorded at least one indictment of a woman. The nature of female crime was almost totally confined to stealing (there were forty-three charges of larceny), followed at a great distance by assault, murder, and misdemeanor (six indictments each).

The 1840s provided the first concrete examples – apart from the long-standing district jails and courthouses – of social dysfunctions being reflected in the "built environment." In some of the towns a building was adapted for use as a surrogate poorhouse before 1840, but it was only in 1841 that the first asylum for the mentally ill was established in temporary quarters in the refurbished York jail in Toronto. In practical terms it was accessible only to the central part of Canada West, and nothing comparable was built in other parts of the province for more than fifteen years. The list of admissions for January 1841 to September 1842, appearing in the commissioners' report for 1842, underlines the importance of accessibility (table 4.3).

Admissions by country of birth reflect the immigration streams: Ireland 68, England 36, Canada 11, Scotland 8, and United States 3. The inauguration of a dedicated facility, although a sign of things to come, did not result in a noticeable amelioration of conditions for the mentally ill, partly because of the unsatisfactory nature of the building and partly because of deficiencies in the abilities of the staff. A permanent building was started in 1845 and occupied in January 1850; but only with the appointment of a progressive medical superintendent in 1854 did conditions for the mentally ill begin the long process of modernization.

Social dysfunctions began to be treated in a less makeshift way in the 1840s. An infrastructure appropriate for a society based upon industry and international commerce began to be put in place, bringing Canada West in line with the world with which it was increasingly interacting. By the end of the 1840s Canada West's population was close to one million. Social problems were outstripping the informal arrangements resorted to in a largely rural farming population. Change was being confronted in gradually more professional ways, and the groundwork for ameliorating many of the social dysfunctions was being laid.

INTENSITY OF SOCIAL STRUCTURING

The vignettes of social functions and dysfunctions offered above have been designed to fill in parts of the picture of the formation of Ontario's modern social fabric, primarily in the form of built facilities. What kind of a social landscape had taken shape as a result of the profound stirrings of the decade or so leading up to 1850? This section attempts to provide a measure of socialization based upon distributions of relevant expressions of society at work.

Where were the "highs" and "lows" of the social landscape of Canada West in the 1840s? The inference in figure 3.3c is that the heart-

Table 4.3
Admissions to the Toronto asylum,
January 1841 to September 1842

Source location	*Number of Admissions*
City of Toronto	59
Home District	18
Newcastle District	10
Gore District	10
Niagara District	7
Johnstown District	6
Midland District	5
Bathurst District	2
Western District	2
Eastern District	1
Wellington District	1
Brock District	1
London District	1
Huron District	1
Prince Edward District	1
Canada East	1
Total	126

Source: Canada, *Journals of Legislative Assembly*, 1842

land areas, with a relatively heavy and evenly spread population, would have been well provided with social facilities, such as schools, post offices, physicians, agricultural organizations, and church buildings, whereas the fringe would have had very few, if any. One might expect a correlation with the density of population and the duration of settlement. A more specific measurement of differences from place to place, however, can only be taken through surrogates, here called social indicators. In this section various indicators presented earlier in the chapter are combined to arrive at a measurement of the "intensity of social structuring."

Intensity is defined as the number of social indicators found in a ten-mile-square block in a grid superimposed over the whole of the southern part of the province. The ten-mile-square dimensions are close to the average size of a township; in fact, this was one of the standard township sizes proposed when surveying began in the 1780s. Intensity was measured for each of the 340 blocks in the grid by adding up the number of occurrences of the seven chosen indicators: agricultural societies, mechanics' institutes, Orange Lodges, temperance societies, medical doctors, post offices, and innovations. Although they do not form a complete picture, these seven indicators do provide a diverse sample of largely spontaneous initiatives by the people living in a block – a series of windows on the "cambium layer" of society. Furthermore, these indicators were the most general examples available; the whole of the province was a potential field of endeavour for each of them.

Agricultural societies were an outgrowth of a well-developed farming settlement (to get a government grant, an agricultural society was required to raise a matching sum), and mechanics' institutes offered a somewhat comparable level of organization for urban workers. The Orange Order, although extremely exclusive in its adherence to Protestant Christianity, was by far the most widespread fraternal and social organization in the province. Temperance societies were patronized by a somewhat different and more general sample of the population. The free choice of location by a medical doctor would normally be a reflection on the settlement conditions in the chosen place. The registration of patents or innovations is accepted as a meaningful indicator of a relatively sophisticated level of development. Post offices, representing the most fundamental contemporary means of communication, constitute a basic "surface" of social organization; there were 480 post offices listed in 1849,[62] and their distribution among the blocks of the grid ranged from zero to five.

The results of the measurement of intensity are presented in figure 4.10. The greatest amount of social structuring is found along the Lake Ontario littoral. Across the province the major urban centres stand out, except for Bytown on the Ottawa River. The retarding effect of the edge of the Canadian Shield can be discerned, and the large blank "hole" of the unsettled Queen's Bush dominates the northwestern corner. The impending opening of the Queen's Bush is signalled, however, by a modest development of social facilities around its perimeter and along a couple of colonization routes north towards Georgian Bay. There are a number of other areas, especially adjacent to Lake Erie, where, forty or more years after the opening of settlement, the social fabric is still thin. A large proportion of the Western District, between the Thames River and Lakes St Clair and Huron, is at a very rudimentary stage of organization. At the eastern end of the province, thirty years of settlement has yielded only five or six of the indicators in

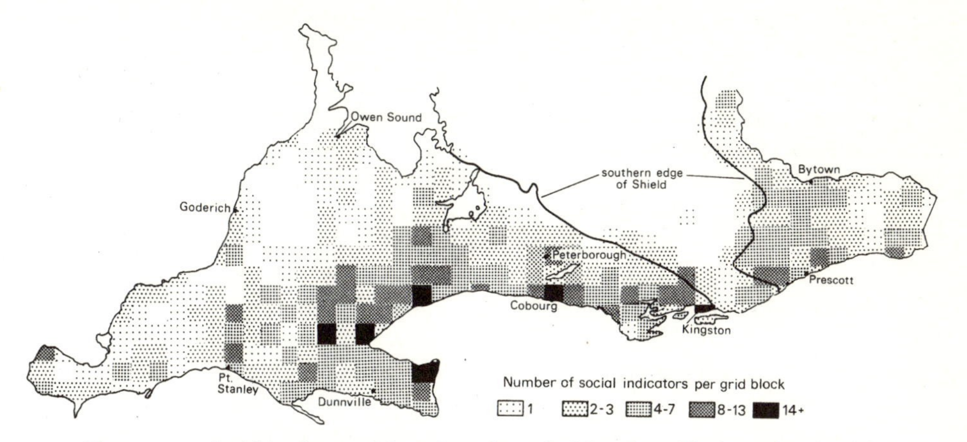

Figure 4.10 Social landscape of Ontario at the end of the 1840s. The intensity of social facilities, based on seven indicators: agricultural society, mechanics' institute, Orange Lodge, temperance society, medical doctor, inventor, post office

many of the townships. A backward section northeast of Prescott remains very poorly served.

A number of corridors into the interior show up in the social fabric, including that from the head of Lake Ontario to the middle Grand River; Port Stanley to London; an embryonic line through Canada Company land to Goderich; Yonge Street north of Toronto; and the road past Rice Lake to Peterborough. The social fabric was attached to, and gained strength from, the complex of connections that reached out toward the fringes of settlement. The connections were not found everywhere, and so their influence varied from place to place. Yonge Street appears to have encouraged the establishment of social facilities toward Lake Simcoe. Roads on both sides of the Grand River probably aided in the development of that corridor. The Welland Canal shows its impact in the Niagara Peninsula pattern, as does the Rideau Canal between Kingston and Bytown (Ottawa). London was connected by a road to the Talbot settlement and Port Stanley on Lake Erie. The whole littoral of Lake Ontario and the St Lawrence Valley was served by a major road and a telegraph line.

Comparing the intensity of socialization in 1850 with the representation of heartland and hinterland in 1840 (see figure 3.3C) underlines the impressive change that took place in the 1840s. In general the heartland of 1840 was well served in 1850, although in relative terms some readjustment was taking place. For example, Windsor/Sandwich seems to have been surpassing Amherstburgh in the southwestern corner, the London area had developed rapidly, social facilities had pushed well up the Grand River Valley, and Peterborough had become the focal point for an area of rapid change. What is even more noticeable is the vaulting from settlement marginality into social development of many sections of the Lake Huron and Georgian Bay shores, eastern Simcoe County, the central Shield edge, and the upper Ottawa Valley. It is instructive to notice the contrasts from place to place; the pattern in figure 4.10, with its heavy emphasis on the centre, is a break from what might be expected, based upon the number of years since initial settlement. Except for the Niagara Peninsula and the immediate vicinity of Kingston, the longest-settled areas were being surpassed by newer-settled areas based largely on prospering agriculture.

DIMENSIONS OF A SOCIAL GEOGRAPHY

During the 1840s, Canada West progressed from a condition of postrebellion confusion to a condition of rapid extension of social organizations and their facilities. The landscape began to be dotted with buildings erected for specific purposes – school houses, temperance halls, Orange Lodges, post offices – to produce a human-dominated scene, especially in most of the areas settled for forty or more years. Between the routes identified for the proposed railways and the forested frontier of settlement in the north, there were definite contrasts: from ten or more concrete indicators of the entrenchment of society within a ten-mile-square block to only one or none at the frontier.

The measure of social intensity shows, however, that even within areas settled for a number of decades the social fabric could be rather threadbare. The area near the centre of the triangle of eastern Ontario, the margins of the Shield in the centre of the province, the corners of Prince Edward District, the sandy lands south and east of Georgian Bay, the Queen's Bush, the wetlands east and south of Lake St Clair, and a stretch along the western Lake Erie shore – these are all examples of the hesitation of people to commit themselves to long-term social organization because of the difficulty of permanently settling these areas. Some of these areas, such as the wetlands and the Queen's Bush, would be transformed by later developments, but most of the rest would continue to have an irregular, economically uncertain career. Regional differences within southern Ontario, most of which were to prove remarkably persistent, were clearly emerging by the middle of the nineteenth century (see an illustration of this in chapter 5).

The census of 1851–52 indicated a population in the province of nearly one million. Immigration, the lifeblood of the colony in earlier days, had become a major problem, especially with the huge influx of Irish famine refugees after 1846. By the end of the decade, at least one in ten of the inhabitants of Ontario must have been a relatively recent Irish immigrant. As in Great Britain a half century or more earlier, the traditional forms of charity were overwhelmed by the numbers of needy people and the increasing transiency of the population. True dysfunctions in society, such as perennial destitution and mental illness, could not be ameliorated by the traditional forms of social organization that were proliferating – by the clubs, the various societies for "improvement," or the churches. New notions and examples of successful treatment of social dysfunctions began to enter Canada West via the channels opening to the outside world. Reform thinking in the United States and Great Britain inspired modernization of the educational system and, less precipitately, the penal system. Through similar channels came more professional treatment of the mentally ill and support for the poor. During the 1840s, social organizations and their facilities, all of which had a role in shaping the landscape, were hurriedly being initiated or expanded in Ontario in preparation for the coming of the railway and the development of an industrial capacity that was the goal during the rest of the century.

5 Making a Living

> A very general feeling prevails ... even in Canada, that her growth and prosperity are not commensurate with that of the United States, and without any inclination to deny or conceal the rapid progress of our neighbours, it may be well, by a few facts ... to prove how erroneous such an impression is.
>
> *Census of 1851–2*[1]

The "few facts" proffered by the official in charge of the 1851–52 census were in the form of an exhaustive disquisition on agricultural crops and population growth. This report showed that Canada West in particular had increased in population faster than even the well-endowed states south of the Great Lakes. In certain agricultural activities Canada West was more productive than neighbouring states. "Upper Canada ... produces six bushels more Wheat per individual than Ohio – the latter producing in her staple Indian Corn twenty-nine times more than Canada, which produces 77 times more Peas, and 54 per cent. more Oats than Ohio."[2]

Ohio was taken to be a fair comparison in terms of quantity of cultivated land and length of settlement. Competing in agricultural productivity was a contest in which Canada West felt at home, because agriculture was the unrivalled livelihood in the province at mid-century. It is not surprising, therefore, that Agriculture was one of the two divisions of the census of 1851–52 (the other being Personal); but it might be considered odd that the census contained no questions specifically related to timber exploitation (discussed below as "The Other Economy"). The agricultural schedule of the census recorded landowner's name, address (i.e., lot and concession), the number of acres held, the division of the land into garden or orchard, total acreages under field crops and under pasture, and amount of wooded or wild land. It then assessed the acreage and production of specific crops, the amount of wool and textiles, butter, cheese, maple sugar, and the domestic stock owned and the yield of animal products.[3] Whereas the personal enumeration had forty-one columns, the agricultural had fifty-five. This kind of information makes it possible to prepare an exceedingly detailed atlas of the province or of a certain district at this date. That agriculture dominated provincial livelihoods, directly or indirectly, is also suggested by the 1842 census, where there were specific questions only about farm servants, domestics, and "persons engaged in trade or commerce" (questions sixty-four to sixty-seven), in addition to the general question about a person's occupation ("trade or profession," question eight).

AGRICULTURE AS THE WAY OF LIFE

The Average Settler

Agricultural settlement was the aim of the province from the beginning, but the aim and the achievement were separated by years of struggle and perseverance. The province was almost totally wooded at the outset. There were a few enclaves with only a scatter of trees, called oak plains or oak openings, that were probably a result of repeated burning by the pre-European occupants.[4] Most settlers avoided the openings, suspecting them of being sterile, and only experienced North American pioneers took advantage of them at first (see the Dumfries Township illustration later in this chapter). There were many other variations in the tree cover, as befits an organic system, including postfire thickets, blowdowns, and different-aged stands of trees.[5] These provided a variety of challenges for clearing. The overwhelming task of the first few years of settlement for the majority of settlers was clearing away the trees and ground cover. Research in recent years has shed more light on the average settler's progress in clearing (rather than that of those who, through reminiscence or fantasy or fortunate circumstances, could boast of rapid clearing). Even the relatively modest rate of two to five acres cleared per year, widely accepted a few years ago, has been refined by Peter Russell to something closer to a mean of one and one-half acres per man per year.[6] With additional hands, and especially with a team of oxen, clearing could proceed at a faster pace; but it is clear that an isolated settler could spend the better part of a working life clearing enough land to have a viable commercial farm. The massive challenge of clearing the woodland has diminished over time in the story of overall pioneer success, but it was *the* factor – usually combined with sickness, particularly the malarial "fever and ague," and isolation – that broke the settlers who gave up.

The term "cleared," in the frontier context, would have applied to a scene far from "clean": in fact, a cleared farm would have included stumps and a great deal of piled and half-burned debris. But it was land on which certain seeds could be broadcast or otherwise planted, especially wheat or corn (maize). In an ambitious survey of the increase of a farm's value through farming practices in the United States, Martin L. Primack calculates that land clearing was the major task of farm improvement even as late as 1900. It accounted for more farm labour than fencing, ditching, and raising farm buildings combined.[7] The rate of clearing in the eastern woodlands of the United States was close to that of southern Ontario, where pioneers faced a similar kind of forest cover.

There were at least two distinct methods of clearing, one more typical of the southern states, the other of the northern states and Ontario. The southern method was characterized by "girdling" the large trees; that is, chopping a band through the cambium layer low on the trunk (called "ring barking" in Australia, where it was used widely). This caused the trees to die quickly and drop the leaf canopy. Over time the trees would lose branches and eventually blow down, if they were not chopped down first. The northern method involved the complete chopping down of all trees and at least an initial assembly of the debris into piles that would eventually be burned. The chopping of the trunks into ten- to fourteen-foot lengths, for hauling to the piles by a team of oxen, was called "junking" in Ontario.[8] Preparing the piles for burning gave rise to the activity called a log-rolling bee. Both methods were known throughout eastern North America, and girdling was tried in Ontario, but in the north it was generally viewed as a debased method, perhaps because of a more urgent desire for crop production, or because of a different image of the ideal farm. The hazards of girdling were underlined in this letter of 1832; the practice could cause injury or even death to humans and domestic animals when the rotted trees began to disintegrate: "the man that was on the lot before I got it had girdled the trees on the greater part of the improvement[,] that is cutting through the bark a little from the root and clearing away the underwood and small trees and in a dry time setting fire to the brush ... the trees when begun to rot is apt to come down in high wind and break fences and kill cattle and when they do stand a length of time they get very hard to cut."[9] The advantage of girdling, on the other hand, was that it required less initial labour to make the land available for some rudimentary farming uses. A thorough review of pioneer technology by Norman R. Ball demonstrates that it was not unusual for both girdling and chopping to be used by a settler, the former especially on trees larger than three feet in diameter.[10]

Primack compares the two methods of clearing by looking at the declared costs per acre. Figures from North Carolina in the 1890s, which were a little less than the common costs fifty years earlier, show that preparing one acre through girdling would take a labourer thirteen and a half days, while chopping would require seventeen to twenty days. This, however, did not include later costs for clearing the dead trees from the girdled acres.[11] Good representative data from Michigan indicate that in the middle of the nineteenth century the costs varied from ten dollars per acre for chopping during the winter, with rough piling for burning a year or two later, to fourteen or fifteen dollars per acre for spring chopping *and* clearing of the debris. This was the equivalent of one man's labour for, one and one and a half months, respectively.[12] Evidence suggests that these costs were close to the going rate in most parts of Ontario at the same time. Either method left the stumps, which could take up to twenty years to rot. Pulling the stumps was as arduous a task as chopping and burning. In the absence of anything more than fragments of evidence, Primack estimates that removing the stumps required only slightly less labour than the initial clearing, although technical improvements eased this task in the second half of the century. To put the clearing process back into the realm of real experience, however, it is well to remember that calculations of hired labour costs do not reflect the physical (and perhaps spiritual) costs exacted from all the pioneers who personally opened land in the forest. Recall John Muir's reflections, probably in Simcoe County, on new settlers grubbing away, "black as demons ... So many acres chopped is their motto."[13]

The diaries of Benjamin Smith provide a remarkable record of a successful life in agriculture, starting with clearing his farm.[14] He settled at the beginning of the century near Ancaster, just west of the Head of the Lake (Ontario). At that time Smith's "agricultural team" consisted of himself and a few seasonal helpers. An analysis of his diary for 1805 suggests that wheat was by far the premier crop, based on the number of references and the types of activities in which it was involved; a sizable part of the winter was devoted to threshing and cleaning grain. All told, Smith seemed to have thirteen or fourteen farm products (see table 5.1). During the same year he spent a great deal of time grubbing and burning trees, and chopping wood for fuel. His harvests were notably late, beginning at the end of September and not finishing (at least for root crops) until November, probably because of more slowly maturing varieties and other demands on his time (see figure 5.1). There were a number of occasions when Smith helped neighbours and built on his own farm. And as with all farmers, the winter was the chief time for killing meat animals and transporting products to mill or market. By 1826, with a number of sons, Smith's farm team amounted to eight or nine men, and in the harvest season about seven other helpers. Wheat was still in a category by itself, but there were sixteen other products that received attention during the year.

Table 5.1
Relative importance of crops and animals on different farms, 1805–45[a]

Benjamin Smith			
1805, pioneer farm nr Ancaster, Gore District	1826, partially developed farm	1837, well-developed farm	
Farm team: farmer + seasonal	Farm team: 8 or 9 men + ca 7 seasonal	Farm team: infirm farmer, 6 men + 2–3 seasonal	Order of importance[b]
Wheat	Wheat	Wheat Wood (mainly lumber)	I
Corn Oats Buckwheat Turnips Flax	Hay Corn Apples Buckwheat Flax Peas Pigs/Pork	Oats Potatoes Apples Hay Barley	II
Fuel wood Orchard Pigs/Pork Beef Potatoes Pumpkins Wool	Fuel wood (or lumber) Oats Maple Sugar Potatoes Beef Rye Wool Cucumbers Cherries	Buckwheat Peas Maize Pigs/Pork Cherries Beef	III

Benjamin Freure	*Benjamin Crawford*	*John Clark*	
1837, pioneer farm, Eramosa Township, Wellington District	1837, partially developed farm, near Beachville, London District	1845, well-developed farm, Ernestown Township, Midland District	
Farm team: ca 3 men + 2 seasonal	Farm team: ca 4 men	Farm team: ca 3–4 + 2–3 seasonal	Order of importance[b]
Wheat Potatoes Hay Wild game (for food)	Wheat Pigs/Pork Barley	Wheat	I
Pigs/Pork Peas Oats Turnips Barley Oxen	Potatoes Hay Peas Oats Maple Syrup/Sugar Buckwheat Flax Turnips Potash ("black salts") Hens	Pork Potatoes Rye	II
Beans Cabbage Raspberries	Wool Corn Beans Pumpkin Cucumber Squash Apples Cherries Strawberries Plums Deer (wild)	Peas Wool Apples Fuel wood Corn Buckwheat Cabbage Maple sugar Cherries Beef	III

Sources: Benjamin Smith diaries (Archives of Ontario); Benjamin Freure diary (Toronto Reference Library); Benjamin Crawford diary (Archives of Ontario); John Clark diary (primarily weather and social reports) (National Archives, MG 24 I 149, vol. 1)

[a] The farm or agricultural team was the group engaged in the long-term working of the farm. At times, workers were employed simultaneously in different activities.

[b] The order of importance of an item is based on the frequency of its appearance in the record of daily farm activities and on the amount of time it occupied in the farm regimen. In every instance wheat was in the first order of importance, whatever the level of development of the farm.

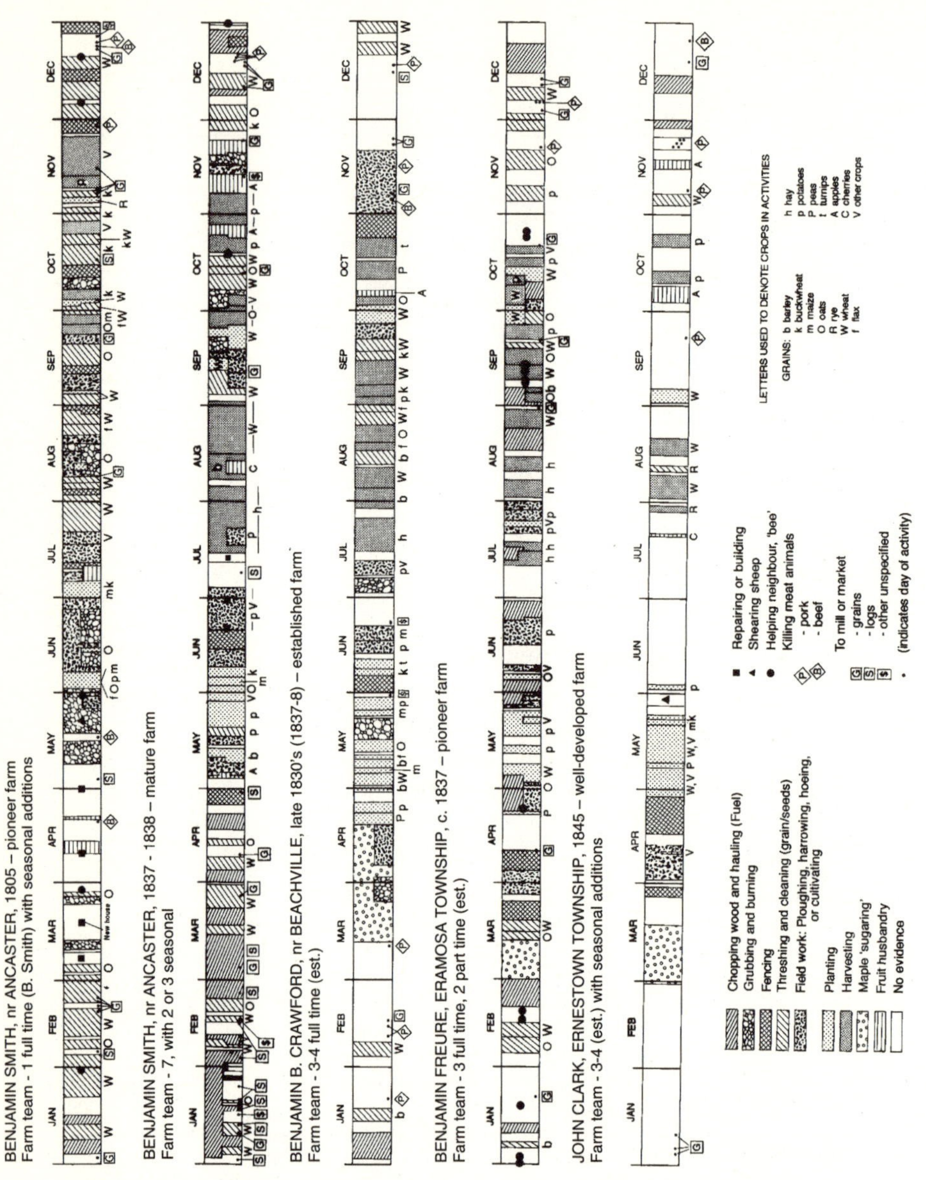

Figure 5.1 The farm year of four farmers: various years (1805–45), locations, and stages (Archives of Ontario [AO], diaries of Benjamin Smith and Benjamin Crawford; Toronto Reference Library, Baldwin Room, diary of Benjamin Freure; National Archives of Canada [NA], MG 24, I 149, vol. 1, diary of John Clark)

By 1837 Smith was well along in his farming career, and may have been beginning to trim down his farming activity to reduce the physical demands. In any case a simpler routine seems to have evolved, with going to the mills, chopping wood, and threshing and cleaning seeds being the winter activities. Ploughing, harrowing, planting, and cultivating occupied workers in early summer. Hay harvesting took place in July, barley and wheat harvesting in August, oats in September, and potatoes in October and November. The number of products receiving attention was back down to approximately the number with which Smith began thirty-five years earlier. His agricultural team, however, contained six men (some of whom might have had a farm of their own) in addition to himself and two or three seasonal additions. Wheat still held pride of place among his crops, but running a close second was the cutting of trees, apparently for lumber. Seventeen trips were made to the saw mill with logs, and although Smith was doing some building at the time, it is possible that some lumber was being sold for building in Hamilton or further afield.

In 1837 Benjamin Freure was almost at the same pioneer stage that Benjamin Smith had been at in 1805, although a little land had already been cleared on the property he bought in Eramosa Township in the new Wellington District.[15] Freure's farming activity at this time focused on wheat, potatoes, and hay, with periodic sorties by one or two of his sons to hunt wild game (a food staple of settlers in the bush). Freure's 1837 diary lists slightly different selection but the same number of products as Smith's diary of 1805 (see table 5.1, above). Freure's agricultural team was largely composed of the young men of his own family. He was probably surrounded by many other farmers engaged as he was in clearing forest and growing basic crops, since his farm was located near the frontier of settlement at this time. This is reflected in the large number of times Freure's farm team helped a neighbour (see his farm year, figure 5.1, above). The winter was the time for cleaning seed, chopping wood, killing pigs, and going to the grist mill. The spring was occupied with "sugaring off," fencing, and ploughing. Freure planted wheat both in spring and fall, and his harvest season stretched from July (hay), through August and September (wheat, oats), to October (root crops).

The third representative farmer – Benjamin Crawford – had a well-developed farm by 1837. He had come originally from New Brunswick, and bought partially cleared land near Beachville, halfway between Woodstock and Ingersoll. His diary suggests a considerable production of wheat, barley, and pork, but in addition to these basic products of Ontario farming in the 1830s, Crawford had a very large range of farming interests, from wheat to wool and cherries, yielding in all two dozen different products.[16] His agricultural team consisted of three or four sons in addition to himself. After the maple syrup collection and field preparations during March and April, the Crawford farm engaged in planting vegetables and grains from late April through May. From late May to the beginning of July, the main activities were grubbing and burning woodland, fencing, and cultivating sprouting crops. In the middle of July, harvesting and threshing began, with hay followed by barley, wheat, and oats. The farm season ended with lifting root crops, fencing, ploughing, slaughtering meat animals, and travelling to the mills.

Eight years later, on a well-developed farm east of the Bay of Quinte, John Clark was focusing on wheat.[17] This was probably after the common wheat pests had passed through the area, but perhaps as a continuing precaution Clark was planting both winter and spring wheat, and a little club wheat.[18] His diary dwells mainly on weather and social events and regrettably little on the farm and farm workers. He was a lieutenant colonel in the militia, a justice of the peace, and apparently a wealthy man, so it is presumed that his agricultural team was one of the premier ones in the district. Ernestown was an unusually fertile township for the Midland District, the northern two-thirds of which were dominated by the rock and shallow soils of the Canadian Shield. Thus, it is less surprising that Clark's farm diverged from the relatively heavy reliance in Midland District on rye and oats. He may also have been making his farm regimen simpler, because he was fifty-eight years old. There are fourteen products that appear in his diary (see table 5.1 and figure 5.1, above).

Like most of the farm diarists, Clark prepared the year's meat supply in November and December (when it was possible to freeze part of it, rather than preserve it in brine), tapped his maple trees in the spring, planted from the end of April to the beginning of June, sheared his sheep at the end of May, harvested from the end of July (hay) through August (winter wheat) and on to October (spring wheat, other grains, fruit, root crops), cleaned seeds and chopped wood during winter, and went by sleigh to the mills when the snow and ice were suitable.[19] Other common winter activities included cutting down and burning trees in land clearing – which was part of the routine even on well-developed farms – building or repairing fences, and hunting. These farmers all produced wheat and pork, generally among their most important products, and they had many other products in common.

The farmers who grew winter wheat planted it during September or early October and reaped it the following August; spring wheat was planted in late April or in May and reaped from mid-September into October. When one stands back from the historic Canadian debate over wheat as an export "staple," it can be seen that this grain was important everywhere in local economies in early settlement in Ontario. Wheat was a wonderfully tractable food crop that could be easily sown and harvested in new clearings. But within twelve or fifteen years the role of wheat had to be re-assessed because infections and insect pests spread with settlement, and the climate or soils of certain areas were not well suited to it. As shown in the regional differentiation later in this chapter, some areas came to rely more on rye or oats. Oats, planted in May and harvested from late September through October, were an essential part of the diet of working horses. It is notable that turnips, which were so popular in improved agriculture in Britain, had made an appearance in Ontario, but carrots are nowhere to be seen in these diaries. Evidence from probate inventories in the northern United States suggests that, from the end of the eighteenth century, the diet of ordinary families began to have a greater variety and quantity of vegetables, and to have less seasonality such as had resulted in springtime shortages.[20]

Farmers like Smith, Crawford, Freure, and Clark were engaged in what Serge Courville has neatly described as "domesticated spaces" (as opposed to the more regimented pattern of the rectangular survey).[21] They were building up both social connections with neighbours, some of whom might become relatives, and geographical relationships with their own properties and districts. These were the beginnings of identification with a landscape, a process celebrated recently by Simon Schama as "topography ... elaborated, and enriched as a homeland."[22] But, as the next section reveals, not all landowners in early Ontario identified with the land in the same way.

Absentee Land Ownership

A notorious irritant that compounded the hardships of initial settling was the persistently empty land set aside in reserves of one kind or another, or owned by absentees. The profound and widespread dislike of the Crown and clergy reserves is well known – expressed even as early as 1817, by Robert Gourlay. The Canada Company was organized to take over the Crown reserves in 1826 and to sell those sevenths of the land on the market. Various schemes were initiated to reduce the amount of land sequestered in the clergy reserves, but they were still being sold after the middle of the century. The land of absentee owners was liberally dotted across the province and remained generally a nuisance.[23]

About the time the debate over the Crown reserves was reaching a climax, a number of new townships were surveyed west of Lake

Simcoe. These townships allowed settlement to spread north of the large tract (eventually Peel and Halton Counties) acquired from the Mississaugas in 1818. It appears that, at least in the surveyor general's office, the main access road through the tract was also expected to become the main road for the region of the new townships and further north. It was to the west and east of this road, called Hurontario Street, that two or three tiers of townships were laid out. In these townships the concessions, running south-north, were numbered from Hurontario Street; for example, south of modern Highway 9, Caledon and Erin Townships; and north of Highway 9, Mono, Adjala, Mulmur, Tosorontio, and Essa on the east side, and Amaranth and Melancthon townships on the west. This numbering from Hurontario Street survives in current survey descriptions.

One of the townships east of Hurontario Street was called Essa. This was a typical rectangular township, surveyed in 1820 into rectangular two-hundred-acre lots, each divided equally into two "square hundreds." (For example, see the squares in the northeast quadrant in Essa Township, figure 5.2.) Essa had reserves, a number of lots taken by special claimants, nearly 5 per cent given as payment to the surveyor, and a portion available for regular settlers.[24] Except for the unusually large number of military claims, coming hard on the heels of the War of 1812, this township was typical of the kinds of activities in landtaking that had characterized the settlement of Ontario since the 1780s. Initially 60 per cent of the township was held by absentees, primarily sons or daughters of United Empire Loyalists (UEL), or military claimants, and 28 per cent was in Crown and clergy reserves. The 12 per cent that was available to bona fide settlers was largely pushed to far reaches of the township by the land of non-residents, displaying an extreme example of the isolating pattern that had been a bone of contention at least since the Simcoe era (figure 5.2).[25]

Essa Township was settled irregularly over the three decades after 1820 and provides a good example of the problems raised by unoccupied land (most of that in reserves or special claims). Settlement spread from the main road allowances on the south and east sides (i.e., from the Yonge Street more than the Hurontario Street side). Because of the relative scarcity of free grant land, some early settlers purchased land from the absentee owners. The free grant land was "free" to bona fide settlers in the sense that they could receive a lot at no cost once they had paid modest administrative fees. After the War of 1812 attempts were made to collect the fees more efficiently, and by 1827 colonial administrators had decided to sell all Crown land by auction in future. The lots available to bona fide settlers in the west and northeast of Essa Township were excessively sandy or poorly drained, and many of them

ESSA TOWNSHIP
(SIMCOE DISTRICT)

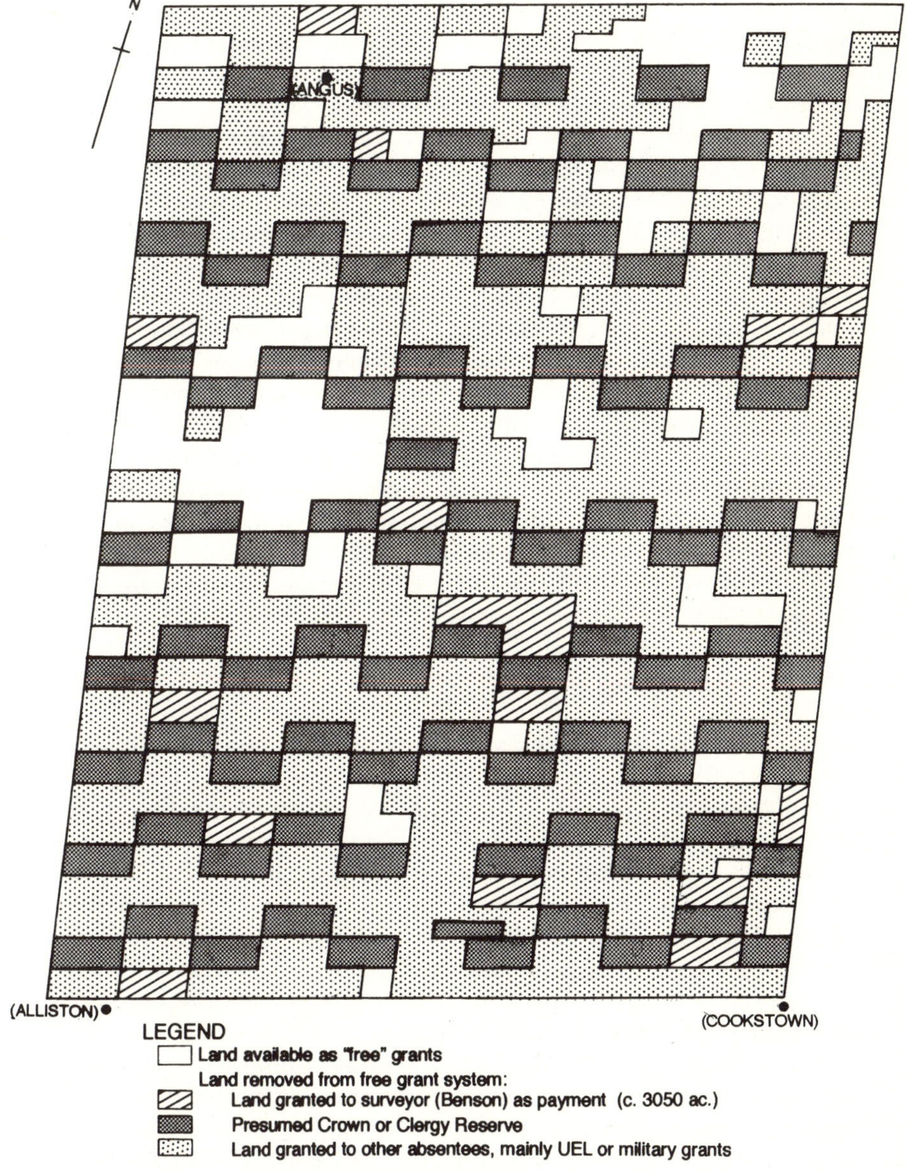

Figure 5.2 Initial allocation of land in Essa Township, northern Home District (Simcoe District, 1837), ca 1821 (Simcoe County Archives, Abstract Index to Deeds)

were not occupied until mid-century. The Abstract Index to Deeds suggests that some of the land open to bona fide settlers was still not purchased as late as 1870; and in the county atlas of 1881, the road grid in the northwest corner was virtually non-existent.[26] As far as the early settlers in a township such as Essa were concerned, their clearings were for many years like beacons in a night sky dominated by the dark mass of uncleared forest, most of it belonging to absentee owners.

It was difficult enough to carry on agriculture in the forest without the added burden of unoccupied and uncleared environs; therefore, it is not surprising that many embryonic farming enterprises foundered. We have no record of the failures, but an oblique measure might be found in the record of land tax arrears. In 1826 the legislative assembly passed an act that would force owners of land on which taxes had not been paid for at least eight years to either pay or forfeit land to the value of the arrears. Lists of the relevant properties, by lot, concession, and township, were published in the *Upper Canada Gazette* in 1828 to 1830. The display of the proportions of tax-delinquent land in the townships across the province (in figure 5.3) offers a geographical vignette of disjuncture, if not failure, in the process of occupying the land of early Ontario. It seems rather incredible that a number of townships had over half their area seriously in arrears, and that a majority of the townships had over one-fifth in arrears. This is not the picture of a homogeneous, well-integrated agricultural land.

The Germination of Ontario's Regions

In the formative years of the agriculture-based economy in Ontario there were significant ups and downs related largely to foreign grain prices. After a difficult period in the late 1830s, a recovery during the 1840s, combined with improvements in roads and canals, put prosperity within the reach of any ambitious farmer. Apart from being the predominant basis for the livelihood of the colony, agriculture also began to rival timber from time to time as the most important component in the export economy.[27] Loss of the assisted entry of Canadian wheat to the British market was balanced or cancelled out by a vigorous growth of export to the United States.[28]

The realization of farm ownership, however, continued to elude a larger number of hopeful pioneers than the frontier myth has suggested. A telling revelation comes from the incomplete census of 1848, which compared numbers of rural landowners and tenants by district.[29] It is wise to treat the census data collected in the 1840s with caution, 1848 certainly being no exception; even up to the 1852 census there is no assurance that the enumerators all worked from a

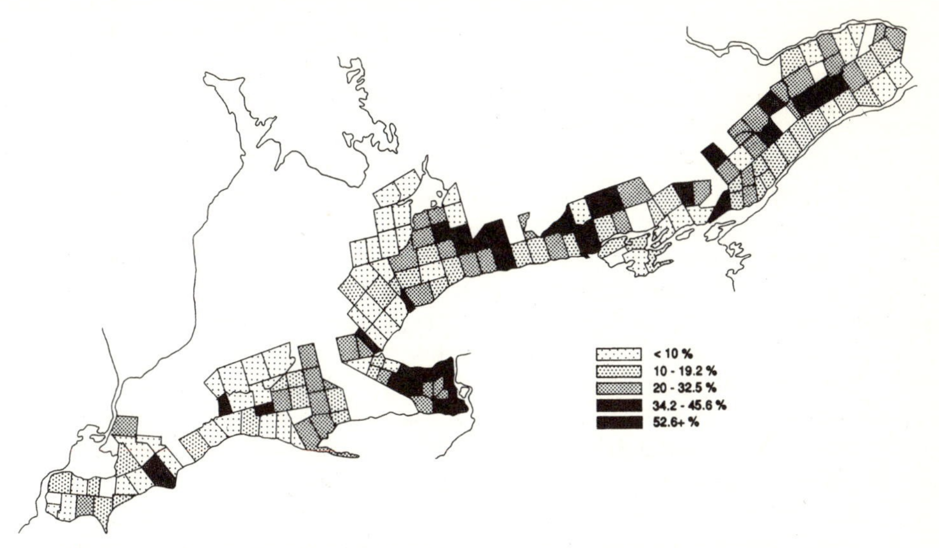

Figure 5.3 Proportion of land declared in tax arrears for eight or more years by township, ca 1830 (*Upper Canada Gazette*, various issues, Sept. 1828 to Feb. 1830). It is possible that data for some new townships (e.g., west of Lake Simcoe) were incomplete.

common understanding of the responsibilities of their occupation. With that in mind, we might note that in three of the twenty districts in 1848 – London, Talbot, and Ottawa – the reports indicate more rural tenants than owners, even though it appears that almost any size of property, even ten acres, was enough to qualify for the landowner category.[30] Working out the figures for the province as a whole yields proportions of 57 per cent "proprietors" (owners) and 43 per cent tenants of rural properties. Surprisingly, other districts that had more tenants than the provincial mean included Niagara, Gore, Home, and Newcastle – that is, most of the oldest and most desirable parts of the province. Districts with high proportions of proprietors were Simcoe, Wellington, and Western; that is, fringe areas (except for parts of Western) into which farm settlers had advanced only in the previous twenty-five years.

Whatever the degree of reliability one accepts for the 1848 census, it is significant that certain aspects of the pattern were maintained through 1871. According to the census of that year, which specifically tallied owners and tenants, the mean proportion of tenants across the province, omitting districts opened to settlement since 1848, was 16.7 per cent. The highest rates of tenancy were found in the richer agricultural areas, such as west of the head of Lake Ontario and along its north shore. There was a notable concentration of tenants surrounding

Toronto, and in two counties east and west of it, where tenancy averaged 25 per cent. Niagara, too, registered a similarly high rate.[31] Though the rates vary between 1848 and 1871, the pattern is similar. It appears that Ontario was still living under the shadow of the pork-barrel land granting to government favourites that dated back to the beginning of the century. This practice kept blocks of the first alienated land effectively out of reach of ordinary settlers and gave rise to the higher rates of land renting.

In many cases tenancy was a rung on the so-called agricultural ladder. A farmer's son might rent a farm in the home area with the hope of earning enough to buy land. Another route to farm – or at least land – ownership was to work as hired help on a farm or in the woods. Robert Leslie Jones claims that wages were not high for farm labourers (and figure 5.7 bears this out) but that an ambitious young man might expect to save enough in four or five years to buy land.[32] A commonly heard complaint was that good farm labourers, as well as good domestic help, were almost impossible to find, but this did not seem to translate into higher wages.

The impediments of absentee owners and reserved land, however, were being overcome in most parts of the province by the 1840s. The clearing away of forest cover gathered momentum, and many of the townships along the shore of Lake Ontario and into the Niagara Peninsula came to be over 40 per cent cleared by mid-century. A majority of the cleared acres would have been under crop. Much of the farm production was for use in British North America, but the surplus was beginning to overtake forest products as the major export. In the older areas, agriculture had been under way for a half century, and metamorphosis was well advanced by the 1840s: the various regions of the province were adapting to the bounty and to the limitations of nature, as well as to the developing lines of trade.

Wheat had undoubtedly been the mainstay of pioneer agriculture right across the province, but over time farmers adapted their practices more subtly to their particular soils, climate, and location, as well as to the advancing crop diseases and pests. Different husbandries were beginning to create different landscapes. The province was no longer Upper Canada the colony, everywhere engrossed with cutting and burning woodland and struggling to rise above subsistence. Some areas, especially those with a large wheat surplus, were key players in the export market to the United States. A start had been made in certain privileged places on specialty crops for which these regions were later to be famous – fruit in Niagara and tobacco in the southwest. Flax was promoted by the government and was grown in substantial amounts in some areas. It became obvious that the eastern third of the province was not well suited to wheat, whereas the western littoral of Lake Ontario seemed to be ideal. A distinct wheat belt materialized, with its apex northwest of Toronto in the recently settled rich lands of Peel and Halton, where the highest per acre production in the Canadas was to be found – a mean of twenty-six and one-half bushels per acre in Esquesing Township (Halton County) in 1851.[33] Another premier wheat growing area was located west of Lake Ontario, stretching toward London. The eastern parts of the province had adapted to environmental realities by putting more effort into oats, rye, and potatoes, some of which probably helped to provision the Ottawa Valley timber extraction. The areas distant from the larger urban centres and market connections tended to fit the classic Von Thunen model and to rely, at an above-average rate, on fattening meat animals.

Because agriculture was so universally important in settled Ontario, the predominant agricultural activities came to characterize different parts of the province, laying the foundations for the regional variations that persist to the present.[34] The agricultural regions (shown in figure 5.4) gradually increased their distinctiveness over time because of the powerful connection between the natural environmental characteristics and the efficiency of crop production demanded by the market. Obviously the boundaries between regions were somewhat porous and flexible, but farmers in a given region would move toward producing the crop or product mix that brought the most profit. By the end of the 1840s the agricultural regions were beginning to crystallize, although the newly opened areas would continue to evolve as they passed through the pioneering phase.

The Ottawa Valley region was dominated by the river as a power source and transport medium of particular importance to the extraction of timber. Much of the agriculture was aimed at supplying the timber businesses with meat, oats, and horses. (This relationship continued to be strong wherever timber exploitation moved, and as late as 1880 it was claimed that "the farmers ... rely almost wholly on the lumbermen for the sale of farm produce" in much of Simcoe County.)[35] In fact, the Ottawa Valley had many limitations for field crops, and agriculture was at a rudimentary stage. Less than 20 per cent of the occupied land was in crops. There was more than half as much land in potatoes as in wheat, and as much land was in oats as in wheat.

The Lower St Lawrence region was one of the oldest settled areas in the province, having received its first agricultural settlers over half a century before. Land quality was very mixed, and farmers had attempted various crops. Oats had proven to be promising in this region too, and thus occupied as much land as wheat. Only 25 per cent of

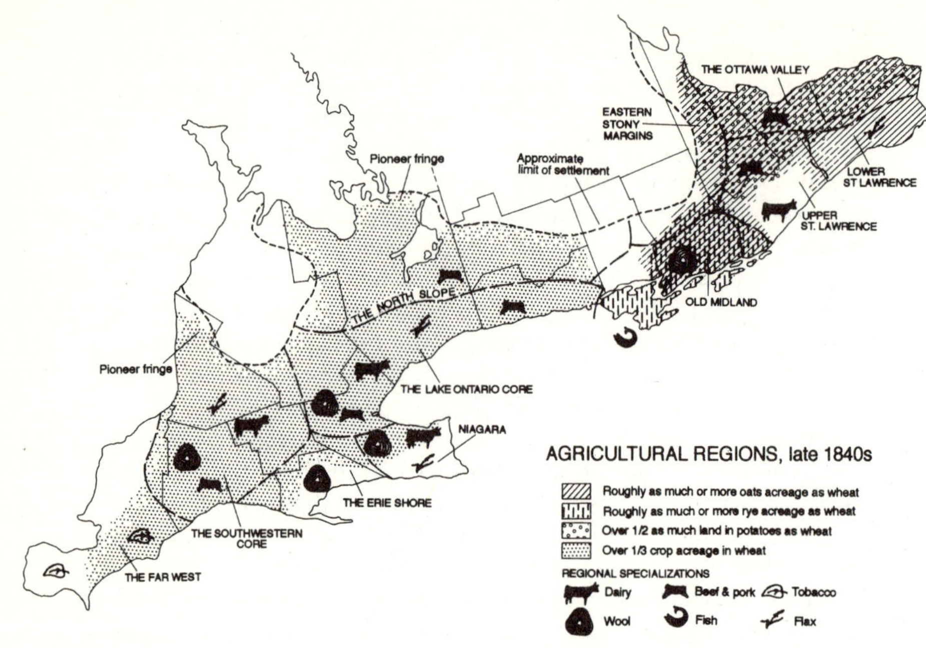

Figure 5.4 Regions of Ontario, late 1840s, based on the dominant economy – agriculture (census data for 1848, in Canada, *Census of 1870–71*, vol. 4; crop acreages calculated from average yields of 1840s)

alienated land was under crops. The Upper St Lawrence was also long settled and had already begun to develop an emphasis on dairying, which led a few years later to the introduction of cheese factories. Further inland, potatoes and oats and the raising of livestock were particularly important, as in the Ottawa Valley and on the shallow soils of the Eastern Stony Margins abutting the Canadian Shield. (See figure 5.4.) The Old Midland region had found that wheat was not ideally suited to its environment and had moved toward a considerable reliance on oats and rye. The North Slope was an area still largely in pioneer conditions, and cropland characteristically was tied to the pioneer staple – wheat. Sales of farm products, however, were mainly in livestock. The Lake Ontario and Southwestern Core regions were areas of singularly successful agriculture where wheat was in its element and where soils and climate combined to encourage a range of products. The dairy emphasis that would become the pride of both areas at a later date was already beginning to take shape. Niagara and the Erie Shore were areas where corn had been a relatively important grain since the opening of settlement in the 1780s. The Erie Shore was one of the five regions producing a significant share of the province's wool, and Niagara also was notable for a few other products, foreshadowing its reputation as a garden. Agricultural success in the Far West region was impeded in many places by poor drainage, but the relatively mild climate and long growing season of this area, while beneficial for crops in general, proved particularly suitable for tobacco, a budding specialty crop. Change would engulf Ontario agriculture in the future, as a result of the railway, widening market opportunities, international competition, and crop pests, but it would primarily take the form of variations on the regional themes established by the 1840s. The only regions that would notably improve their relative economic status were those by the major railway lines, between the Niagara and Detroit Rivers.[36]

Apart from being by far the largest economic sector ("class" in census terminology) in the province, agriculture was also responsible for most of the other economic activity. The "industrial class," for example, was engaged in flour milling, blacksmithing, manufacturing farm equipment, saw milling in areas of land clearing, refining fibres and making textiles, distilling, and brewing. Many other occupations, such as shoemaker, tailor, mason, and teacher, were tributary to the agricultural population. Through all the vicissitudes of agriculture, such as the difficult period from the mid-1830s to the early 1840s, there was one consistent theme – a shortage of agricultural labourers. In a period of growth, in 1846, the *Toronto Mirror* observed: "we scarcely speak to a single farmer who comes into the City, that is not crying out for harvest labourers ... The poorest hands, quite inexperienced in the work of the Country, refuse to go for less than fifteen dollars a month and board."[37]

If the evidence of the census of 1848 is to be relied upon, the shortage was extreme. There were nearly 58,000 proprietors of rural real estate and 43,000 rural tenants, most of whom would have been in farming. At the same time there were only 7,866 farm labourers listed, and 3,483 male servants (many of whom would have been employed on farms).[38] A rough calculation would suggest that, on average, only one farm in ten would have had "a hired hand." This could only be believed if many of the proprietors held very small farms – ten acres or less, the minimum suggested by the census of 1851–52. In a quaint reiteration by the *Toronto Mirror* one week later, we learn that "labourers continue to be sought after with unabated solicitude, and are not to be had."[39]

THE OTHER ECONOMY: TIMBER

Up to the 1840s Ontario's export economy had been based on the timber business, whereas the domestic economy for almost all ordinary citizens had been connected to agriculture. Real timber exploitation through the extraction of large numbers of logs by teams of "shanty-men" was an activity beyond settlement, far away from most Ontarians, although trees were being turned into fuel and lumber right across the province.[40] The number of saw mills is a measure of the steady deforestation that was going on. The censuses of 1840 and 1848 show that saw mills were on the increase everywhere during that decade, except in the Niagara and Western districts (table 5.2).

One suspects that the increase in saw mills reflects the rapidly growing demand for lumber in the towns of Ontario and, more significantly, in the mushrooming cities across the Great Lakes. But for Ontario and Quebec combined, according to Arthur Lower's calculations, forest products accounted for just over 55 per cent of the value of exports in 1849, dropping to 50 per cent in 1850.[41] Most of these were still going to Britain, although the amount going to the United States was to catch up in the 1860s. Because Ontario was an even heavier exporter of agricultural products – especially wheat and flour – than Quebec (Lower Canada), it seems likely that agriculture had become the major source of exports from Ontario for some years in the 1840s.

It is possible to compare the strengths of agriculture and forest products by looking at the records of the harbours across the province. At Port Sarnia, at the western extremity, forest products substantially outweighed agricultural ones in 1844, while Port Sarnia's neighbour, Windsor, exported almost exclusively agricultural products (see table 5.3). The only forest products exported from Port Stanley on Lake Erie were two hundred or so kegs and barrels of ashes, and a modest board footage of walnut lumber. Hamilton, at the west end of Lake Ontario, was exporting both agricultural and forest products in 1844, but agricultural products were highest in value. Port Credit was exporting a considerable amount of forest products – indeed, almost as much lumber as Toronto – but, in value, agriculture was paramount, mainly because of the flour. Toronto exported over one and one-half million board feet of lumber, which suggests that Toronto, like Port Sarnia, Port Credit, Windsor Harbour (Whitby), and Napanee, still had a hinterland that was being vigorously cleared. As the main port, Toronto had a wide range of goods in export, and by far the greatest value came from agriculture. In fact, it appears that none of the exporting ports in William Smith's gazetteer – except Port Sarnia, tiny Dawn (tapping the same "back country"), and Napanee – showed a greater value from forest products than from agriculture. This fits with Ontario's reputation for relying on its agricultural prowess.

None of the export reports in Smith's gazetteer were from the far eastern part of Ontario, however, and there was no export table for

Table 5.2
Number of saw mills, by district

District	*1840*	*1848*
Western	16	17
London	73	186
Talbot	37	126
Brock	42	65
Niagara	94	90
Gore	157	148
Wellington	(in Gore)	69
Huron	–	33
Simcoe	(in Home)	38
Home	201	275
Newcastle	98	135
Colborne	(in Newcastle)	28
Prince Edward	38	49
Victoria	25	44
Midland	51	65
Johnstown	57	69
Eastern	39	57
Ottawa	19	26
Dalhousie	–	18
Bathurst	37	46

Source: Canada, *Census of 1870–71*, 4: 131, 171

Kingston and Belleville. Meanwhile, the "other economy" of Ontario, carried on at a distance from most residents, was represented by the large quantities of squared timber floating past Bytown on the Ottawa River from the main source areas for export timber at this time – the upper reaches of the watershed.[42] The Ottawa Valley and Belleville, both of which tapped the Canadian Shield forests, would have tipped the balance in the direction of forest products (see figure 5.5A). The vast watershed of the Ottawa River was not well endowed for agriculture. It had large expanses of poor drainage and a crazy-quilt pattern of soil types; toward the edges of the basin, the overburden was shallow over limestone country rock, and the Canadian Shield occupied the periphery. Most of the parts of the watershed, however, supported tree growth, including some of the most magnificent stands of white pine

Table 5.3
Exports from selected smaller ports in Ontario, 1844[a]

	Port Sarnia		*Windsor*		*Port Stanley*	
	Amount[b]	*Value (in £)*	*Amount*[b]	*Value (in £)*	*Amount*[b]	*Value (in £)*
Lumber	120,000 bf	180	–		63,272 bf	158
Staves	2,800	31.5	–		–	
Potash/ashes	400 b	2,000	61 b	290	74 k 119 b	634
Wheat	300 bu	56	4,642 bu	696	23,186 bu	4,066
Flour	–		65 b	65	4,984 b	5,147.5
Beef	10 b	17.5	98 t	294	200 b	300
Fish	1,000 b	1,250	93 b	163	–	
Sugar	12 b	24	–		–	
Pork	–		434 b	1,164	504 b	1,008
Lard	–		77 b 226 k		17 b	42.5
Furs/skins	–		3 b		166 b	498
Hams	–		70 t 22 h		–	
Bacon	–		4 boxes	16	–	
Tongues	–		15 k	22.5	–	
Butter	–		–		77 b	120.5
Barley	–		–		1,108 bu	111
Grass seeds	–		–		52 b	26
Peas	–		–		2 b	7s. 6d.
Cranberries	–		–		8 b	5
Tobacco	–		33 k	99	–	–
Whiskey	–		–		17 b	25.5
Rags	–		–		Bulk	58

Source: Smith, *Canadian Gazetteer*, 35, 147, 152, 165, 221

[a] Cobourg exported ashes, lumber, staves (West Indies and standard), shingles, beef, pot barley, flour, meal, pork, peas, butter, bran, and liquors. Picton exported ashes, wheat, flour, pork, peas, barley, rye, buckwheat, Indian corn, "coarse grain," fish, leather, and butter. Smith records most values as estimated; some are rounded for this table

[b] b=barrels; bu=bushels; bf=board feet; k=keg; t=tierce; h=hogshead

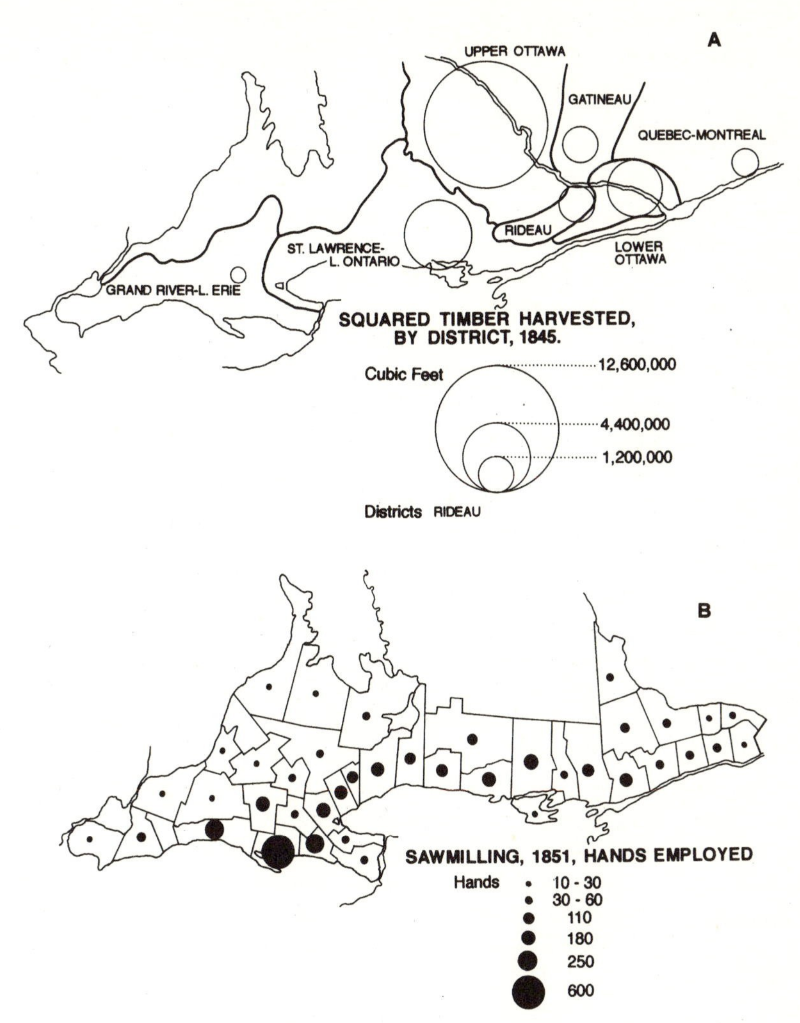

Figure 5.5 Locational dichotomy of timber exploitation: (A) squared; (B) sawn (Head, "Forest Exploitation," figs. 6.7, 6.11)

in eastern North America. The Ottawa had convenient outlets to the east coast, via the St Lawrence River, and it became a major player in the export trade. The Ottawa River watershed was famous for timber – especially the huge white pine and oak "sticks" (squared in the woods with the broad axe) that became important for building British naval vessels during the Napoleonic Wars. The Ottawa watershed retained this distinction for fine timber, but other Shield watersheds also gained prominence as the nineteenth century progressed.[43]

By the 1840s a dichotomy in the exploitation of the forests was taking shape: squared timber was the dominant product from the woods in more remote areas with good watercourses for conducting logs to the St Lawrence for sale overseas; saw logging was the dominant activity in areas closer to the Great Lakes, where ravenous markets for lumber were not far away, along or across the lake (compare parts A and B of figure 5.5). The truly massive extraction was focused on the areas where timber shanty crews were taking out all the large trees, and usually squaring them, for export down the rivers – as huge rafts – and overseas. The other form of timber exploitation was carried out at the opposite extreme from the upper reaches of the watersheds – mainly near the lakeshores, in the thousands of saw mills that had been set up in the areas settled for agriculture. The workforce in the saw mills was more evenly spread across the southern districts; timber was commonly brought in from the clearing of land in the vicinity, and the major product was lumber for local, regional, or even more distant building. Figure 5.5B illustrates the relatively even spread of saw mill workers by 1851, but also shows a notable concentration west of the head of Lake Ontario, in what had been parts of London, Talbot and Brock Districts.[44] This important industry, largely for markets in New York State, was based on a unique forest of superior pine growing on an extensive sandy delta deposited by a glacial river.

In part, the timber dichotomy was based on differences in the forest. On the Shield, conifers, including the versatile white pine that was in demand in Europe, were the predominant cover. Moving south toward the Great Lakes, deciduous hardwoods became more and more common until, approaching Lake Erie, a complex deciduous association was the norm (excluding the aforementioned pine forest). Reconstructing the forest cover on the eve of settlement in the Grand River Valley provides a picture probably valid for much of the earliest settled districts adjacent to the Lakes (figure 5.6). In Dumfries Township there were both conifers and hardwoods, depending on soil type and perhaps previous human intervention. The hardwoods accounted for about two-thirds of the tree cover (not including the "thickets" which were remnants of blowdowns or fires), dominating most of the wooded land south of Galt.[45] The hardwoods had to share the township with a surface type more extensive than conifers – the distinctive "oak openings" or "oak plains." These openings were common in southwestern Ontario, in southern Michigan, and in Ohio, becoming larger and larger to the west until giving way to prairie proper.

There were certain fundamental relationships between trees and sites. It was quite clear that spruce and cedar would usually inhabit damp sites, whereas pine was found on sandy, well-drained land. The

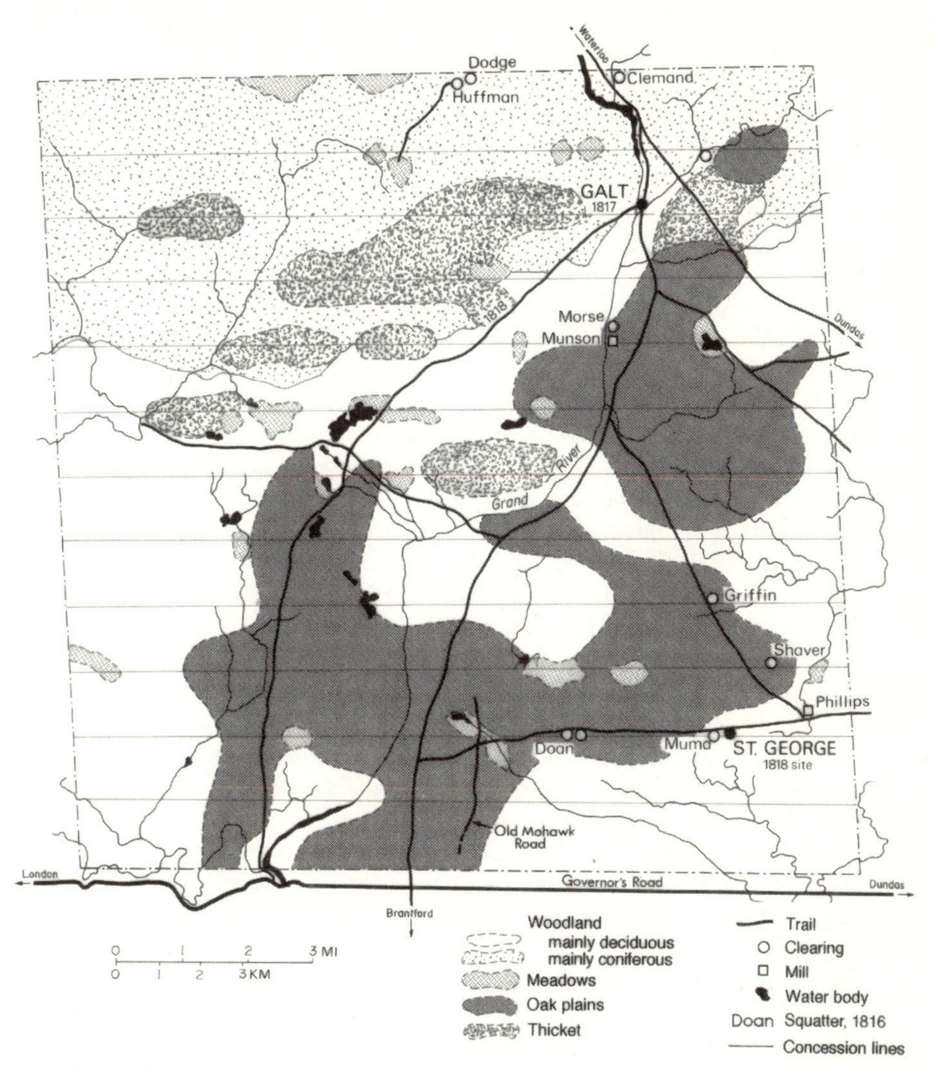

Figure 5.6 Setting the first settlers found in Dumfries Township, 1817–25 (AO, Field notebook (1816–17) of surveyor Adrian Marlett)

hardwoods did not have such clear relationships, except that they seemed to grow on soils generally usable for agriculture. Nineteenth-century "experts" offered many versions of the "fertility by tree type" formula, which settlers often took as gospel, but the complexity of the forest and of the regolith, and the common settlers' ignorance of North American conditions, meant that no formula could claim success. What was important for the beginning farmer was that most of the land was well-drained and that there was a road nearby that gave access to supplies, a market, and other settlers.[46]

It should be mentioned that there were some reliable experts in the judging of land. They were usually born in North America and were often the so-called professional pioneers who partially cleared a farm, then sold it and moved on. Sometimes they also squatted temporarily, as in Dumfries Township (figure 5.6). Squatters did not become a problem in early Ontario, probably for a number of reasons: land costs and settlement duties were light; there was a universal attachment to plough agriculture; the availability of good land did not diminish for over half a century; and there was fairly effective government at the township level.

The first economic encounter between settlers and a standing forest occurred when they arrived at their lots to begin clearing. This was the first step toward creating a farm and ultimately an agricultural economy. Most scholars who have tackled the clearing process have come to the conclusion that earlier estimates of the amount of clearing per year were far above what an average settler would have accomplished. (See the previous section of this chapter.) Differences would arise, of course, from variations in forest composition and the kind of land surface, the number of hands and draft animals, as well as the technique chosen: cutting, piling, burning trees, and grubbing roots (in whole or in part); or girdling, which only postponed the other steps of the process.[47] What were the economics of clearing woodland? How much did it "cost" to clear land? A settler could use up a great deal of "physical capital" in the exhausting process of clearing, especially where large hardwoods were involved. Although the spending of time has been investigated, the very personal costs have not been calculated. From a welter of descriptions of early settlement experiences in Ontario one can determine, for example, that two men with a pair of oxen would have done well to clear (not girdle) four acres of mature wooded land in a year (mostly during the winter), as illustrated in the farm year of farmer Benjamin Smith, figure 5.1. Also, the descriptions indicate that, although women had many duties on farms, the heavy manual labour of clearing was expected to be done by men.

Settlers continued to cut trees even after they had a few acres for farming.[48] Many would take trees to a saw mill to sell for lumber. In this role they would bring together the usually separate worlds of timber and agriculture. From the local mill these sawlogs would eventually move toward market and account for part of the millions of board feet shipped out of the southern ports like Toronto, Port Credit, Windsor (Whitby), and Napanee. Although divergent in activity and primary location, timber and agriculture were sometimes drawn together because of the need for labourers in the winter timber extraction. Some farmers participated in the seasonal migration to work in the woods, cutting, hauling, and driving logs.[49] This could provide a vital supplement to the meagre produce of a bush farm. So too could another product of the forest – ashes, often called the pioneer's first crop. The ashes would be gathered, put in a container and irrigated with water to remove impurities, then dried. The result was a form of potassium carbonate called potash or, if further cleaned and recrystallized, pearl ash. There was obviously a demand for this material: one barrel of potash was valued at four times a barrel of flour, three times a barrel of whisky, and two times a barrel of pork.[50] Something could also be gained by gathering oak bark for use in tanning hides.

Timber exploitation was one of the two pillars that supported Ontario's economy in the 1840s, though the shanties and river runs were located mainly toward the outer edges of the ecumene, while agriculture dominated most of the settled area. As with most of the frontiers in the westward movement across North America, working in the woods was a vital income supplement for many settlers on embryonic farms.[51] Timber was also an integral part of the agricultural economy, based on saw mills located as close as possible to where settlers were cutting trees on their bush farms and wanting lumber. (Many of these mills were movable and would disappear after a year or two.) The opportunity to continue to augment income through the sale of saw logs, fuel wood, and potash from the clearing of a farm, likely made the difference for many pioneers between staying on the farm or retreating to the town. But the process of making a living through timber and agriculture took a toll on the land. For example, the 61,000-acre Vaughan Township, just north of Toronto, contained twenty-five saw mills in 1845 and thirty-four in 1851 – an average of one for each 2,000 acres of land. In 1850 47.2 per cent of the township was woodland; by 1910 that figure was reduced to 5.6 per cent.[52]

SMALL SEEDS OF INDUSTRY

The industries in Ontario in the 1840s fell into two fundamental categories: supporting and servicing the agricultural economy, and supporting and servicing the forest exploitation. Until the mid-nineteenth century, there were no grounds for the emergence of industry independent of the two mainstays of the economy. Even the iron mining that developed on a small scale only survived as long as it could sell its product for agricultural machinery. All consumers were dependent ultimately on agriculture or timber for their spending power. Jacob Spelt has demonstrated for southcentral Ontario that even by 1851 there were few large industrial employers: the mean number of workers (usually including the employer) per establishment in a selection of the main industries was 4.1.[53] The only kinds of industries that had averages much higher were woollen factories (9.6) and foundries (14.6). In both cases, these figures were elevated by a few large establishments. There were rays of hope, however, for what Spelt called "propelling" industries – such as foundries and agricultural machinery makers – that would generate a need for subsidiary industries and various services. Access to knowledge from the United States and Britain about the organization of larger industries, such as canal and railway building, gradually introduced workforce complexity far beyond what was required in agriculture.[54] There was a lively inventiveness in the population (as demonstrated in the previous chapter), and with an increasing reservoir of labour, more major industries were on the horizon.

The only industries employing more than half a dozen persons in 1851 were found in a handful of urban places. The largest urban centres, except Toronto, had all acquired foundries in the previous five years – London and Bytown had one each, and Kingston and Hamilton, two each.[55] Toronto was not stagnating: between 1845 and the end of 1851, its population increased by 50 per cent, to 30,775. There were numerous examples of industries with growth potential building on the prevailing agricultural economy. In Woodstock there was "Brown, A.P. & Co., iron founders and manufacturers of steam engines, mill gearing, stoves, copper & tin ware, threshing machines, ploughs, &c."[56] In Brantford, Goold, Bennett and Company were publicized as "iron and brass founders, manufacturers of stoves and stove furniture, mill gearing, steam engine work, agricultural implements, &c., Colborne st." In Kingston there was the Gore St Foundry, "John Honeyman, iron and brass foundry, and steam engine works, potash kettle and stove manufactory"; in Port Hope "Porter, Archibald, iron foundry, manufacturer of ploughs, wagons, stoves, and blacksmithing establishment, Walton st"; in Prescott "Hulbert, Samuel, general foundry, and manufactory of ploughs, steam engines and machinery, Water st." Cobourg had Albert Yerington's "Victoria foundry, College st – manufactures, steam engines, grist and saw mill irons, Helm's patent circular saw mills, and Ruttan's patent Canadian ventilator"; Brockville had "Colton, R.P., iron founder, Water st, manufactures every description of steam engines and other machinery, equal to any other house in Canada"; and Ayr had John Watson's "Ayr Foundry, manufactures stoves, agricultural implements, threshing machines, &c."

There were other embryonic propelling industries, besides foundries, scattered across the province. In Bytown, for example, there was "Humphries & McDougall, Bytown carriage factory, No. 1, Daly st,

Lower Town – every description of vehicles constantly on hand or made to order, equal to any manufactured in Canada"; and in Hamilton, "Hills, A.H., builder, machinist and agricultural implement manufacturer, James st" and "Hamilton Coach and Carriage Factory, King Street. J.M. Williams returns thanks to a liberal and discerning public for the patronage he has received. As he employs 70 hands, and turns out 10 vehicles weekly, he is enabled to fill all orders on the best of terms at the shortest possible notice, for cash or approved notes."[57] Brantford had a significant manufacturer of pottery: "Morton & Co., stoneware factory and pottery, Dalhousie st. – this is the only *stoneware* factory in C.W. [Canada West], and every article will be warranted equal to any made in North America, and be supplied upon as good terms." Brockville had "Skinner & McCullough, manufacturers of hames, saddletrees, scythe snaths, manure and hay forks, grain shovels, &c., Water st"; and Cobourg had one of the biggest industries in Ontario: "McKechnie & Winans, Ontario woollen mills, King st – over 200,000 yards of cloth are manufactured here annually, giving employment to over 170 persons."

Building stone was quarried in various parts of the province, but most notably in the limestone aureole along the southern Shield edge. Limestone from quarries and granite from glacial erratics provided a pleasing ambience and feeling of permanency to urban centres such as Kingston and Perth. (See the photograph of a pre-1850 stone house in Lanark County in the illustration section.) The Niagara Escarpment was also quarried from the early days of settlement, and building stone, from which the urban scenes of Paris, Galt, and Guelph benefitted, was found in the Grand River Valley. An adjunct to building in stone or brick was the burning of limestone to yield lime for mortar. This activity was carried on in many parts of the province where limestone was found. Gypsum, or "plaster," was for a number of decades in the mid-nineteenth century a famous and useful product of the Grand River Valley. Paris was one centre of production and Cayuga, further down the river, was another. The plaster was spread on farm fields as a dressing, and probably gained its high reputation from helping to balance acidic soils formed in glacial tills with a large Precambrian component; as gypsum, it was important in the making of mortar, plaster, and cement. Brickyards sprang up in many parts of the province where suitable clay was found, and an observant eye can distinguish one locale and its clay from another through the varying colour of bricks in old buildings, ranging from light yellow to umber.

Agriculture was directly represented in the grist milling industry, and the number of millstones began to rise very rapidly after 1825. Thomas McIlwraith demonstrates the increase for central Ontario, where during the 1840s alone the number of pairs of millstones rose by roughly 60 per cent.[58] This increase occurred despite the ending of privileged entry to the British market and because of the expansion of a market much closer to home. The primary reason for building a grist mill was to provide flour for a local population, but flour was as easy to ship as wheat, and, as McIlwraith calculates from figures of population and wheat production, there was a large area of flour deficiency southeast of Ontario in the United States.[59] The Erie Canal, and especially its Oswego feeder from Lake Ontario, was a perfect conduit for flour and wheat from Ontario's usual surplus.

The forest-based industries were almost entirely footloose before the emergence of paper made from wood pulp. There were some large saw mills that were permanent fixtures, but most were small, temporary operations. The reason for moving was usually the exhaustion of readily available trees for lumber. The Ottawa Valley was famed for the sheer bulk of its timber output, and it continued to produce primarily squared timber, as did the other forest districts in and around the Shield. Lumber was a different product, from different districts. In the 1840s, as intimated in the previous section, it was focused in the counties along the shores of Lake Ontario and Lake Erie, drawing on tributary areas that were still being cleared for agriculture. Even the lakeshore townships were usually less than half cleared at this time, and the extraction would continue over the next fifty years until many of the townships would be considered overcleared by some contemporaries. The saw milling industry illustrated in figure 5.5B (above) was the medium through which this massive landscape modification was effected.

Many of the blacksmiths in the southwestern peninsula of the province were supplied with iron from the earliest ironworks, at Normandale. This pioneering establishment was set up in 1809 by New York immigrant Joseph Van Norman, based on an extensive deposit of bog ore across five townships on the mainland north of Long Point, Lake Erie.[60] The works prospered for three decades, but eventually succumbed to the importation of generally superior and cheaper foreign iron. A similar deposit of bog iron led to the brief career of an ironworks, during the early 1830s, near the west end of the province, in Gosfield Township.[61] Toward the east end of the province, north of the Bay of Quinte, hard rock mining of iron ore was developed at the southern edge of the Shield, although after a decade or two most workings were reduced to irregular production or abandoned. A major problem seemed to be the haul of thirty miles or more to reach main transportation routes. Marmora had a promising iron mine from 1820, and nearby Madoc was smelting iron at the end of the 1830s and into the 1840s. Part of the roller-coaster career of the Marmora Iron Works was directed by the pioneer Van Norman, who refitted the furnace in 1847. This recovery only lasted a few years, until much cheaper, imported iron undercut the Marmora product.[62]

The canals came to serve as foci for embryonic industries, especially those that required shipping of their products. The records of the canals and harbours also give us an idea of what kinds of manufacturing were beginning in the middle 1840s. Other than common agricultural goods (flour, oatmeal, ashes, pork, lard, and butter) the only products with a modicum of manufacturing process to pass through the Desjardins Canal at the head of Lake Ontario in 1845 were whisky, quarried stone, and barrel staves, none of which was likely to be the foundation of an industrial flowering.[63] In 1844 the nearby port of Hamilton shipped out 6,121 hundredweight of "Domestic Manufactures" and some stone passed through the Burlington Bay Canal, but nothing else to expand the list from the Desjardins Canal, which primarily served neighbouring Dundas. At Kingston there were shipyard and boat-repair facilities. In the list of exports from Toronto, the only new items of simple manufacture were starch, shingles, and leather; and from the Humber River, woollen cloth. All the other exporting ports were handling only agricultural or forest products, minimally modified if at all.

This seems a rather flimsy base on which to build an industrial economy, but apparently there was potential, because industry was developing robustly before the end of the century.[64] The potential being nurtured in the minds of entrepreneurs and inventors in most parts of the province was to be expedited by the harnessing of steam power, not only in the form of railways, but also in the form of movable steam engines. The latter were already appearing in Ontario in the 1840s, and local foundries were building them. Their revolutionary role was to free industry from the restrictions of river valley locations, and to put the power source nearer the raw materials or the market. The general distribution of steam-powered mills in 1852 is one of the components shown in figure 5.7.

In addition to flexibility of location, another advantage of steam power was its consistency, compared to the increasingly variable flow in the streams whose watersheds were being cleared of original vegetation. Steam engines, however, did not suddenly oust water power: as figure 5.7 reveals, steam-powered mills were to be found in many areas where water power would continue to be a viable source for large mills, such as in the Grand River Valley. There is no apparent connection between the assessed affluence of a township and the attraction to steam power. Concentrations of steam mills are to be found in the most affluent townships, but also in the poorer ones. Although the actual loca-

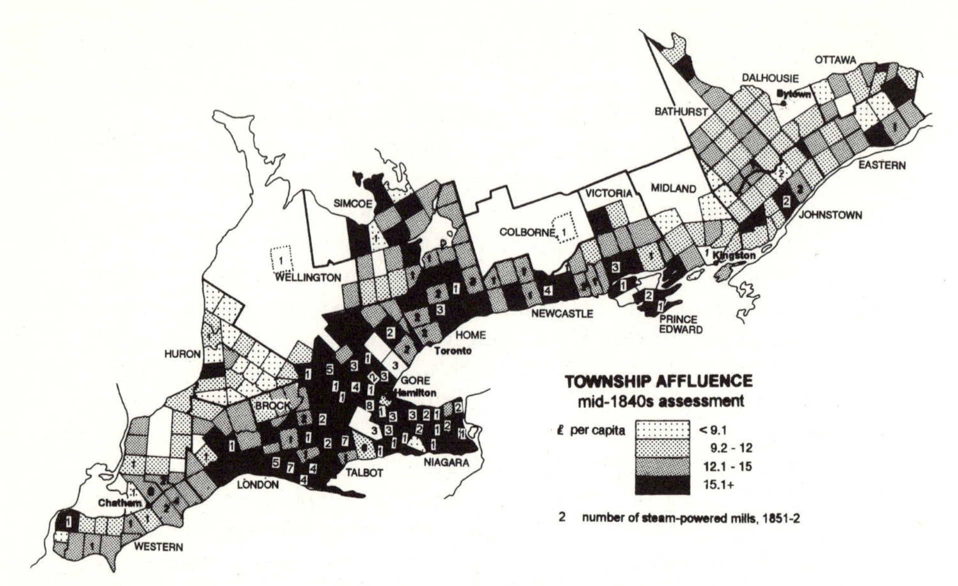

Figure 5.7 Variation in affluence, by township, mid-1840s (Smith, *Canadian Gazetteer,* esp. his comments on assessment, 245; steam mills in Canada, Board of Registration and Statistics, *Census of 1851–2*, vol. 2, no. 7). New townships with steam mills by 1852 are outlined with a broken line; districts and some towns are named.

tions of the steam-powered mills is not known, they seem to have been moving into areas where most of the journey from the foundry of manufacture to the site of operation could have been made by water or on ice, which was probably an important consideration with a large and heavy machine. As a further refinement, it might be noted that there was almost no steam power in use north and east of Kingston, or north of a great arch from Kingston to the south shore of Lake Simcoe to Windsor.

VARIATION IN AFFLUENCE AND SOCIO-ECONOMIC STATUS

In the 1840s Canada was moving through a political process that led to responsible government. In an odd convergence of desires and needs, influential politicians in the home country and in the colony found themselves coming to similar conclusions about their future relationship. The expense of maintaining certain colonies had for some time outweighed what the home country saw as the advantages. On the other hand, both the British government and the colony could see that – as in the case of the United States – cutting the ties would not lead to economic ruin on either side, but would reduce or eliminate the nuisances of governing from across the Atlantic Ocean.

The colonial way of life, however, was entrenched in the thoughts and actions of many people. The old guard was in the process of losing power, but to a considerable extent the province was still run by a number of regionally powerful patrons, local versions of the Home District's so-called Family Compact.[65] The hierarchy of the Home District was mirrored to a certain extent in many other districts. Distinctions in income, and in the security of income, were taken for granted. There was a gamut of affluence and of concomitant social status. There were few that were very wealthy and many that were very poor, including those that had only seasonal employment. It has been persuasively argued that for most urban families a single income was not enough to maintain a decent standard of living;[66] on the farms everyone worked, including women and children.

Long before the calamitous Irish emigration, there were crowds of destitute people descending on Ontario. The *Toronto Mirror* raised an alarm in January 1841, protesting "against the premature and hot-house forcing of 'starving thousands' up on the country, without first placing the means of earning a subsistence within their reach."[67] Another newspaper reinforced the message a few weeks later in a poem, "Remember the Poor":

The season of gloom has arrived,
And winter is hard at the door,
He whispers to all, "my power is revived,"
And tells us, "remember the poor."
The rich who with plenty are crowned,
Who have an abundance in store,
With liberal hands should be found,
Dispensing relief to the poor.

. . . .

Go visit the sick man in bed,
Or look at the couch on the floor,
His wife and his children no bread –
And then "you'll remember the poor."[68]

The questionnaire sent out by Governor General Lord Sydenham's office in 1840 (discussed in chapter 4) gathered a wide range of information among which were pay scales for a large number of occupations. From district to district there were sometimes substantial differences in remuneration, but these were usually found in less common occupations. Some of the aberrations seem to have arisen from uncertainty over whether or not the pay included board. In many of the occupations security of employment did not exist, and in some of them seasonal layoff was customary. In fact, it seems likely that few "blue-collar" workers would have been employed continuously, because at six shillings, three pence or seven shillings per day (rates commonly indicated in the responses to the questionnaire) they would have earned close to one hundred pounds per year. That would have pushed them too far in the direction of the speaker of the Legislative Assembly, who earned one thousand pounds, or even the clerk, who earned five hundred pounds in 1849.[69]

Some of the wage contrasts, as revealed in the chief emigrant agent's report from Toronto, have been discussed in chapter 4. The differences from district to district in the province can be illustrated by comparing two responses to Lord Sydenham's questionnaire – from the chief emigrant agent (Home District) and from Sheriff Treadwell (Ottawa District). The contrast between a relatively peripheral area (Ottawa) and the "heartland" of the province (Home) is represented by the light and dark shaded bars in figure 5.8. The two districts reported wages for slightly different sets of occupations, with the Ottawa sheriff also providing wages by the month for work that could be long term. A few occupations were reported as year-round employment (figure 5.8, right-hand column); namely, miller, tanner, groom, dairywoman, farm labourer, female cook, butcher, blacksmith, and baker. It seems that wages around Toronto, especially for the building trades, were 20 to 30 per cent higher than in the Ottawa area. In both districts there were large ranges in income, in which farm labourers, grooms, dairywomen, female cooks, and quarrymen were at the bottom end of the scale, and shipwrights, millwrights, and some cabinetmakers were toward the upper end. Tanners seemed to be in greater demand in the Ottawa District, but in the Home District curriers were needed, perhaps reflecting a market for finer leather goods. Most of the other occupations clustered around a mid-range wage. With almost all the occupations there was the possibility that the employer could provide room and board, in which case the wage would be one shilling to one shilling, six pence less per day.

In some cases differences from district to district could have been the result of small numbers in a particular occupation. There were some variations, however, that appear to be meaningful. For example, in Bathurst District, the demands of the major logging activity may explain why butchers earned noticeably more than they did in Home or Ottawa Districts, and a little more than in Wellington. The response from the district of Northumberland, on the north shore of Lake Ontario, provided a useful insight regarding remuneration: "the above are

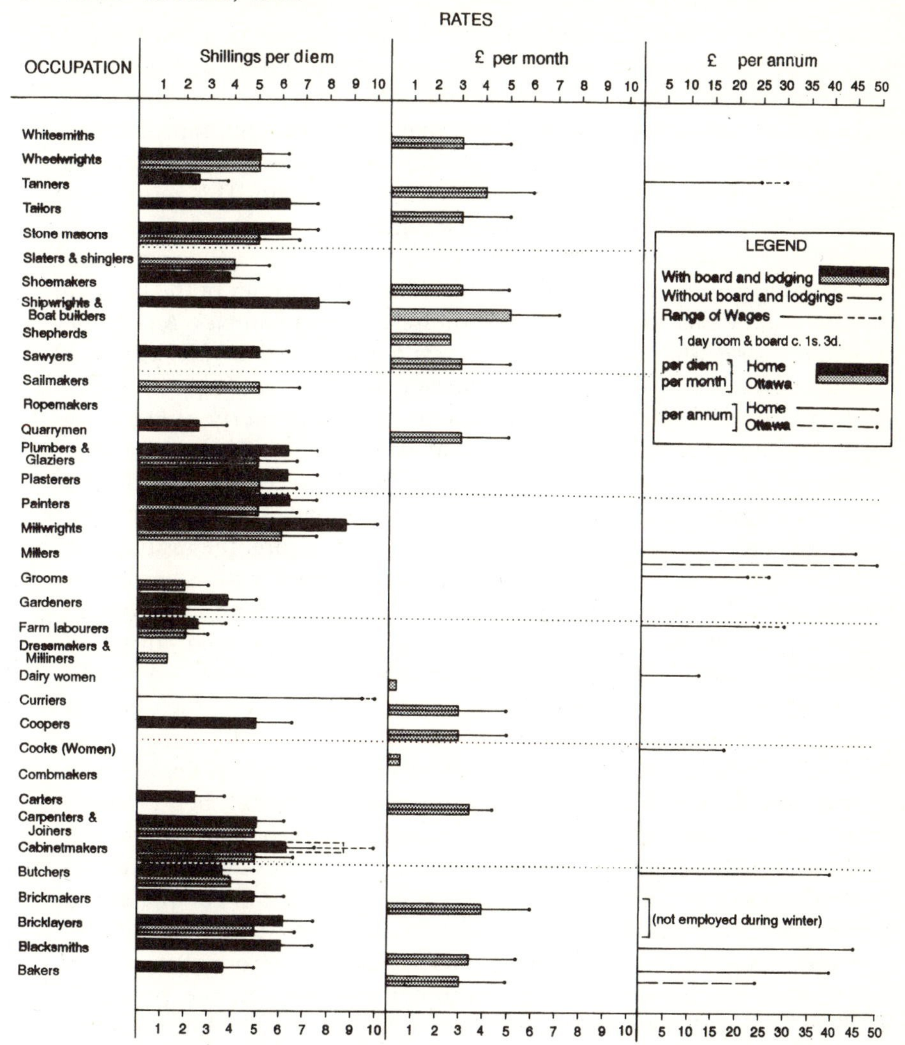

Figure 5.8 Wages paid for a wide range of occupations, Home and Ottawa Districts (NA, RG 5, B 21, vol. 1, pts 1, 2, 1840). At this time, £1 was worth about $4; there were 20 shillings in the pound, and 12 pence in the shilling.

the rates charged hereabouts by master workmen, who when they employ assistants usually pay them a little more than the highest wages given to agricultural labourers. Some master workmen charge more than the above stated rates and in the rural settlements, carpenters, masons etc often work as low as 4s 6d per day without board. Those tradesmen etc whose rates of wages are not stated generally work by the job."[70] Working by the job suggests discontinuous employment. Nineteen of the occupations in Northumberland indicated no rates of pay.

There is evidence that allows us to go on from consideration of a hierarchy of occupations to consideration of general economic variations from place to place. Evidence that there were different levels of affluence across the province comes from the property assessment that was carried out on a regular basis from an early date. This source has been criticized as inaccurate, and certainly there are cases where it is difficult to match names from an assessment with those on a census of the same year. In such cases the considerable transiency of the population is appealed to as an explanation for the discrepancies. A contemporary observer noted that there was a great deal of property not included in the assessment schedule, and also that "as few persons ever return anything like the whole amount of their property, at least five-and-twenty percent. may fairly be added."[71] The justification for making use of the assessment is that it is the only socio-economic evidence, apart from population numbers, that was formally collected from all the occupied townships in the province. Furthermore, the assumption is made that the deficiency in reporting was roughly the same everywhere, and therefore comparisons should be roughly valid. William Smith reports that "the only articles of property assessed consist of land cultivated and uncultivated, houses, water grist and saw mills (steam mills are not assessed), merchants' shops, store-houses, horses, milch-cows, cattle over two years old, distilleries, and pleasure wagons."[72] To give a measure of general affluence in a township, the data have been rendered as so much assessment per capita, then assigned to one of four categories, and prepared as a map (figure 5.7). Although not part of the assessment, the general locations of steam-powered mills in 1852, shown in figure 5.7, were a relevant component in the scene of gainful activity. The population figures in Smith's *Gazetteer* came mainly from 1842, while the assessments were from 1845.

Figure 5.7 shows significant economic contrasts in the province. There was a distinct belt of affluence along the north shore of Lake Ontario, from the Bay of Quinte to the Head of the Lake, with strong extensions into the Niagara Peninsula and westward to take in the Grand River Valley and a broad wedge down to the Lake Erie shore. As discussed earlier in this chapter, the rather unexpected affluence of the counties on the eastern shore of Lake Erie was likely related to the active extraction of lumber, provided by a large number of saw mills that would have been included in a township's assessment. The west end of the province was almost all in the middle range, but the east end had a majority of townships in the two lowest categories. Eastern Ontario was already becoming a region characterized by the exodus of population, after twenty-five to fifty years of struggle with difficult conditions for agriculture. The wedge of the Huron Tract, abutting the Lake Huron shore, is apparent as a district of low assessment. It had only been opened to settlement under the Canada Company for just over a dozen years, and capital investment was at an early stage.

Through various processes, the regions of Ontario were taking on contrasting "personalities" by the 1840s that were to underlie regional differences a century and a half later. The distribution of affluence in figure 5.7 provides a ready reference to fundamental economic differences based on the agricultural variations described earlier (see figure 5.4).[73] The extensively cleared and well-cultivated townships toward the west end of Lake Ontario were gaining affluence based on successful wheat growing, and the Niagara Peninsula was similarly advanced thanks to varied agricultural activities and burgeoning urban markets nearby. On the other hand, much of far eastern Ontario had benefitted from supplying the timber industry in the Ottawa Valley, but the exploitable timber was now at a considerable distance and the demand for supplies had diminished. Figure 5.7 reveals that the southwestern end of the province also had relatively low affluence, perhaps because of the expanses of poorly drained land that would come under major draining only in the next half century, and because it was beyond the regions of rapid population increase and sizable markets.

Where a system of communications was set up to streamline resource exploitation, for example, the modification of streams and building of roads for timber in the Ottawa Valley, the accessibility and links could encourage other activities and support the growth of service centres. The Rideau Canal was an additional underpinning for development in an area that was not inherently ideal for agriculture (but as figure 5.7 shows, even an early infrastructure was not sufficient to maintain widespread affluence). At the opposite end of the province was the contrasting case of the huge Enniskillen Swamp (a major example of the poor drainage in the southwest) that was inherently very rich but had not been approached by any roads. Such an area offered little enticement to settlers and, in the absence of legislative encouragement to drain it, remained a blank area on the map until it was drained later in the century. Thus, the developing communication links, whether road or waterway, could be both cause and effect of the economic status of a particular area. The facilities that allowed for the circulation of goods, people, and information around the province are discussed in chapter 6, and further illustrated in figures 7.2 and 8.1.

6 Circulation of Goods, People, and Information

> In a new Country like this where the means ... are totally inadequate to the Construction of Good Roads, Providence kindly sends us snow and frost to Enable the Back settlers to bring their produce to market, the Lumberman to draw the timber to the streams ... and the farmer to draw his fuel and fencing.
>
> Sheriff Charles P. Treadwell, 1840[1]

On a woodland frontier, getting around was normally difficult. Over time the roads would have improved, although as long as they were dirt, or even gravel, there was no improvement that could long withstand the steadily increasing traffic. The narrow iron-shod wheels on the wagons quickly cut into the surface of even a well-drained and compacted road. During the 1830s the ingenious process of macadamization, in which broken stone was compacted to form a kind of interlocking resistant surface, was introduced to the province. With turnpiking, which included grading and side-ditch drainage, all-season roads began to be extended toward the edges of settlement. The roads, in a sense, were injectors through which the seeds of ecological transformation entered pristine Upper Canada. It is probably recognized better today than during early settlement – witness the struggles over keeping logging roads from invading old forests – that profound change would begin once easy accessibility was provided for farm settlers. A logged, cleared, drained, and cambered road was the ideal in a progressive colony, but even a simple, unadorned trail that allowed easy movement when snow was on the ground opened the way for initial modifications of the wilderness.

GETTING AROUND IN EARLY ONTARIO

The gradual extension of roads, navigable water routes, and communications media allowed early Ontarians to move further from their home bases, to do business, and to pass on information – all constituting a life-sustaining flow for a developing agricultural colonization. The circulation of goods, people, and information was essential to the vitality of both the core and peripheral settlement, as the circulation of nutrients is essential to the vitality of an organism. It is well known that in pioneering days winter was the best time for travelling, because snow and ice reduced the friction of distance. The seasonal difference could be dramatic, ranging from an average of one or two miles an hour through the ruts and corduroy of warm-weather roads to seven to nine miles an hour, often away from the road allowances, by sleigh. Gliding quietly over the snow on a crisp, sunny winter day made that season in the colony more enjoyable and efficient for travelling than in most of the peripheral parts of the British Isles, where macadamizing and bridge building were only being gradually extended. One of the better known encomiums on winter travel in Canada is by Anna Brownell Jameson in January 1837:

> At half-past eight Mr Campbell was at the door in a very pretty commodious sleigh, in form like a barouche, with the head up. I was absolutely buried in furs; a blanket ... was folded round my limbs; buffalo and bear-skins were heaped over all ... thus fortified and accoutred, off we flew, literally "urged by storms along the slippery way," for the weather was terrific.
>
> I think that but for this journey I never could have imagined the sublime desolation of a northern winter, and it has impressed me strongly. In the first place, the whole atmosphere appeared as if converted into snow, which fell in thick, tiny, starry flakes, till the buffalo robes and furs about us appeared like swansdown, and the harness on the horses of the same delicate material. The whole earth was a white waste: the road, on which the sleigh-track was just perceptible, ran for miles in a straight line ... The only living thing I saw in a space of about twenty miles was a magnificent bald eagle, which ... alighted on the topmost branch of a blasted pine, and slowly folding his great wide wings, looked down upon us as we glided beneath him.[2]

It was in the months from November through February that the majority of heavy farm products were transported to market or mill. The agricultural year diagram (figure 5.1) for Benjamin Smith shows that in the winter months of 1805 he made eleven off-farm trips to mill or market. In the winter of 1837–38, he, or a member of his agricultural team, made twenty recorded trips. During the other eight months of 1805, he made only four trips, and only seven in 1837–38 (excluding social visits by horseback). It is easy to imagine how much more convenient it would have been to move wagonloads of grain to the grist mill, logs to the saw mill, and barrels of salt pork or of potash to a merchant's shop, when there was snow on the ground. With the temperature below freezing, wet spots and streams had a solid surface, ruts were buried, insects were non-existent, spillage and spoilage were not a major concern, and short cuts were feasible.

As settlement progressed, statute labour by landowners was applied to improving the roads. An act of 1793 formalized the terms of statute labour for Upper Canada.[3] Widening (meaning cutting more trees and letting in the sunshine), draining of the road allowances, and building simple bridges were the basic accomplishments under the statute labour system. The general improvement and extension of non-military roads was slow because of the difficulty of mobilizing the settlers for their township road duty. The requirement of statute labour, to be applied to building and repairing roads, gradually gave way during the century to the payment of a levy that could be used by the municipality for more effective road work, although even in the 1950s statute labour was still being performed in some parts of Ontario.[4]

The public roads were established on the allowances of the grid survey, and most earlier spontaneous trails were abandoned. A few of the trails became so useful and important that they were retained, despite diverging from the rectangular grid. These so-called "given" roads were of two general types: long-distance pre-European trails, and cross-country access routes to a mill or entrepôt. To a large extent Lt-Gov. John Graves Simcoe's system of military roads, which is still important in today's highway pattern, was of the first type. Among prominent examples of long-distance given roads would be Islington Avenue, linking northwest Toronto to Woodbridge and Kleinburg; Highway 12 from Orillia to Coldwater and west; the East and West River Roads between Galt and Paris on either side of the Grand River (see fig. 5.6); the old highway through the Niagara Peninsula (Highway 8); the highway from Kingston northeast into the Ottawa Valley; and wandering roads in many parts of the province where the land is rough enough to break the relentless survey grid. Countless examples of the second type, including many of the "Mill Streets" of the province, can be found scattered across present-day county road maps of the Ministry of Transportation.

The most highly publicized road innovation of the 1840s was the plank road. The degree of contrast between this wonderfully smooth, new road and the rough, rutted surfaces of even the gravelled roads gave the plank road an aura of fantasy. It was like driving through an endless, narrow ballroom. One editor, as reported by G.P. de T. Glazebrook, thought the horse "seemed more in a gay frolick than at labour."[5] This type of road, like the railroad that came after it, promised to save the dainty traveller from the misery Catharine Parr Traill had suffered a few years earlier, when she "was dreadfully fatigued with this day's travelling, being literally bruised black and blue."[6]

The development of the plank road gave a stimulus to the surveyors and their brethren. Glazebrook records that the first plank road on the continent was laid on the east side of Toronto in 1835–36.[7] Plank roads began to appear in the northern United States from 1840, probably as a result of their popularity in Canada.[8] By the late 1830s, plank roads were being proposed for many parts of Ontario, such as from Lake Ontario to Rice Lake. In 1842 the Board of Works called for tenders for constructing this road, and instructed the civil engineer Nichol Hugh Baird to prepare the specifications. The road was to traverse a variety of conditions and would include sections of standard road allowances that were currently in use. Fences were to be removed and the organic surface horizon scraped back to leave a swath of thirty-six feet, thus providing a wide thoroughfare for rapid passage in both summer and winter. (The specifications and construction technique indicated must have been generally similar to the requirements for any advanced road of the day, such as a turnpike, whether planked or not). The terms of the job, as illustrated in figure 6.1, continued as follows: "excavation from the ditches to form the crown of the road as shown ... to be six inches above the natural surface of road, so that the sleepers when laid shall bed thereon. Formation for road-way to be sixteen feet level on top, and for six feet on each side falling an inch to a foot, from thence curving to bottom of drain as shown on section, allowing two feet bottom of drain level." These were some of the instructions for building on presently travelled roads. For a road allowance that was partially cleared, there were some additional instructions: "in plain formation [i.e., flat] or Embankment not exceeding 5 feet, the road to be grubbed to the width of 36 feet or outside of ditches, to be cleared to the full width of allowance or 66 feet. When the surface corresponds with the grade or twelve inches above, the vegetable soil to be removed from road surface for the width of 18 feet." As for the all-too-common "swamps and swales," they were

> to be grubbed and cleared as described and "slashed" outside of road allowance to a total width of 175 feet. The first operation after clearing off, to be – draining ... In swamps the foundation to be laid with one or more tiers of fascines in bundles well "withied" together, of 18 or 20 inches diameter at the waist in convenient lengths to be laid down so as to break band with each other [staggered rows?]. When these fascines shall reach within 12 inches of required grade, the difference to be covered with the best material to be procured (not vegetable soil) ... these fascines to extend to 24 feet in width of road, with side drains of 18 inches at bottom.[9]

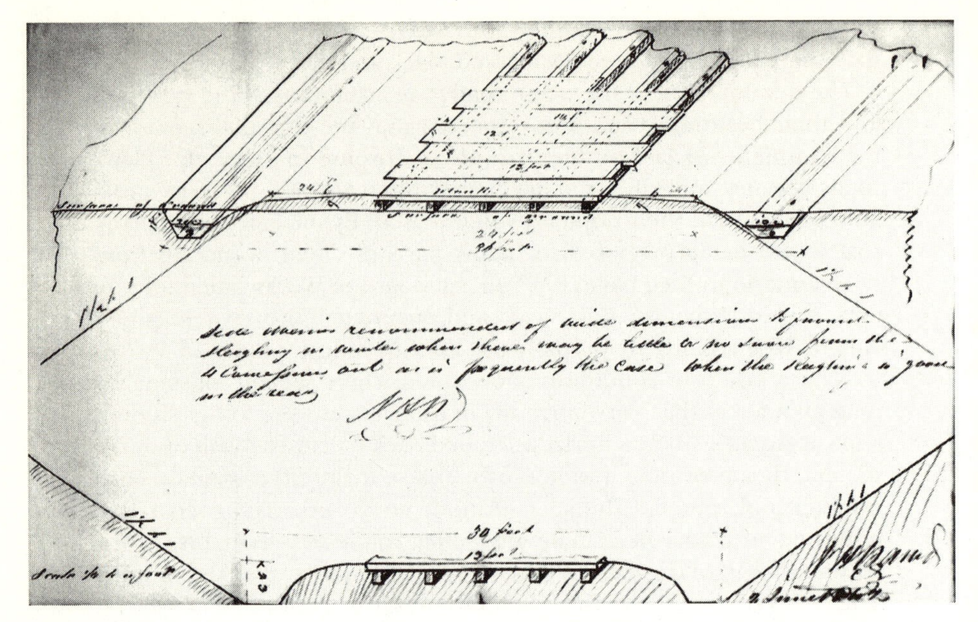

Figure 6.1 Construction diagram for a plank road, by N.H. Baird for Canada Board of Works, 1842, which shows an oblique view above and a cross-section below (National Archives of Canada, RG 11, vol. 81, no. 2)

These specifications were dated June 1842; the foundations and embankments were to be completed by the following November. The twelve-foot-wide plank road, laid crossways on sleepers that were four by six inches square, would be constructed on this bed, and "where the line shall pass along occupied lands, the road [was] to be fenced off with a Cedar Snake fence or oak rails 5 in height, 'stake and rider.'" The note on Baird's diagram (figure 6.1) reads "side drains recommended of wide dimensions to funnel sleighing in winter when there may be little or no snow from the 4 Concession out as is frequently the case when the sleighing is good in the rear." This road appears to have been a successor to the Cobourg and Rice Lake Plank Road of the late 1830s, which in turn had taken up the mission to access the back country from the aborted Cobourg Railroad Company (chartered in 1834, see discussion below).

Initially the building of a plank road was greeted with almost universal rejoicing in the tributary areas. The dramatic improvement in travel during the first couple of years on the plank road almost made the paying of tolls – the usual means of maintaining such a road – acceptable. Plank roads, however, had their detractors from the beginning. The reality of maintaining the structures after the first two or three years gradually deflated the initial excessive optimism and steadily increased the body of doubters. Even the three-inch oak planks that were commonly used deteriorated under the traffic of shod horses and metal-bound wagonwheels. It was not long before planks had to be replaced. Accidents would occur, especially as a result of a hoof plunging through a weakened plank, and maintenance costs and insurance claims could eventually surpass the income generated by the tolls. Traffic congestion developed, especially where the plank road met other kinds of roads. Generally within five years the plank road became an economic liability.

The first plank road in Simcoe County played out the typical saga. It was constructed by the county as an extension of Yonge Street, from Holland Landing to Bond Head, in the expectation that it would more than pay for itself. The road made a profit for the first few years and carried a huge amount of heavy traffic. Bernice Merrick Ellis recreates the scene in *Life on the Old Plank Road*: "laden wagons viewed from the highest point on the road stretched as far as you could see in both directions." But this popularity took its toll: "planks splintered and broke. Repair costs mounted. The number of accidents to wagons, horses and riders increased. In 1857 there was not only no profit from the Plank Road but the County paid out £250 in repairs and £34 for damages to persons injured on the road [The Bradford *Chronicle* asked] 'Do the Commissioners ... intend to repair the road this season? It is now almost completely worn out and nothing doing to it. Yet the Commissioners still modestly demand tolls.'"[10] A few months later, in October 1858, the council voted to dismantle the plank road after seven years of operation.

BEYOND ROAD CONSTRUCTION: MAKING CONNECTIONS

Up to the 1830s most travel and communications moved along roads, trails, and water routes in the lakes and large rivers. Steamboats were operating on the Great Lakes by the 1820s, and by 1830 the canal era had begun. In the 1830s land travel began to improve through the macadamizing or planking of roads. Before this, travellers had regularly commented on the state of the roads, almost invariably in negative terms.[11] But William Smith was able to traverse the province by road in gathering material for his *Canadian Gazetteer* in 1845 (see his road map reproduced in figure 7.2), and for his classic *Canada: Past, Present and Future* in 1850. Through his descriptions he led his readers along the routes travelled, and he provided reference maps that showed the generally passable roads. Most parts of the province were laced with roads of varying quality. As one would expect, the number and the quality of the roads diminished toward the fringes and in poorly settled sections.

All in all, the construction of passable roads pressed ahead greatly during the 1840s. It was at the end of the decade that Baron de Rottenburg, assistant quartermaster general, prepared his justly famous map of the roads. The map he produced was in twelve sheets, totalling about twelve by twenty feet, and the detail it showed for most areas was commensurate with its size. It was entitled "Map of the Principal Communications of Canada West," and it was compiled from over two dozen competent representations of parts of the province.[12] Although the detail varied somewhat from place to place, by and large this work indicated all the passable roads and classified them according to construction (from macadamized or planked to "obscure roads") and condition (such as, "bad," "tolerably good," "bad corduroy") and it reported inns, mills, and urban centres.

As a by-product or a reciprocal development, stagecoach travel rose toward its zenith in the 1840s, to be supplemented and then overwhelmed by railways in succeeding decades. Stagecoach lines were established and timetables were published. For example, Smith's useful gazetteer of the mid-1840s included stage timetables for the larger towns. London had a daily stage to Ingersoll, Woodstock, Brantford, and Hamilton, and in the opposite direction daily to Chatham and Detroit, three times a week to Port Sarnia, and twice a week to Goderich. Another stage went Monday, Wednesday, and Friday south to Port Stanley on a new plank road.[13] Similarly Hamilton had stages to Toronto, Port Dover, St Catharines, Galt, Guelph, and London. St Catharines had regular stagecoach connection to Hamilton, Niagara, Buffalo, Chippawa, Dunnville, and Queenston. Port Hope, Cobourg, and Belleville, being on the main east-west routeway, had the daily Toronto-Kingston stages pass through. Cobourg also had a daily stage that went inland to connect with Peterborough. Remote Goderich, at the far limit of the Canada Company's Huron Tract on Lake Huron, had a stagecoach twice a week to Galt as well as to London. As the transportation hub, Toronto had connections with all parts of the province. There were also local coaches during the day from Toronto to Thornhill, Richmond Hill, Streetsville, and Cooksville.

These road building and transportation initiatives had pushed back the curtain of wilderness and extended the ecumene (or "habitable world"). Using the road system on de Rottenburg's map, and making three miles or less from a road the criterion for habitability (or practical access to the outside world), the ecumene can be shown to occupy most of peninsular Ontario by the end of the 1840s (see figure 6.2).

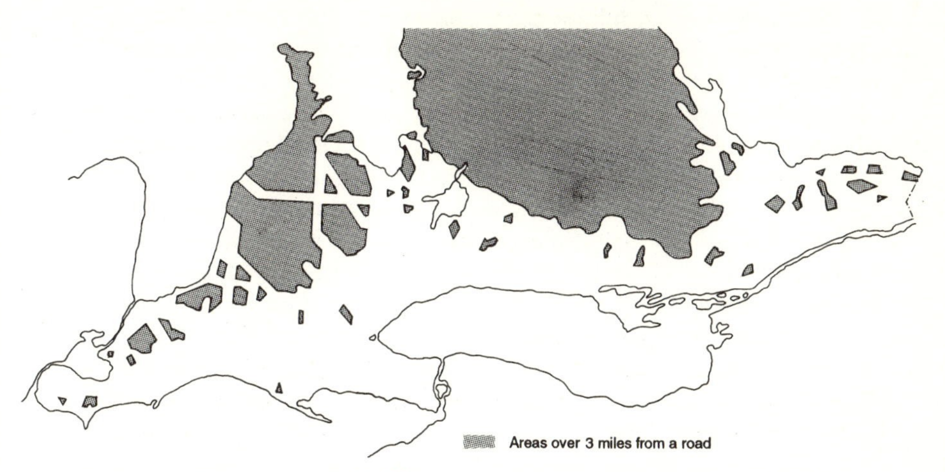

Figure 6.2 The ecumene (habitable area) of southern Ontario, ca 1850 (Whebell, "Robert Baldwin," map 1)

The Canadian Shield to the north, obviously inimical to farm settlement, stands out starkly; in the west the Huron Tract still has some blocks of relative inaccessibility, and incursions by colonization roads have only begun to pierce the Queen's Bush in the northwest. (The full picture of the ecumene *and* the urban places that were supporting the onslaught on the wilderness is presented in chapter 7, on urbanization.)

Complementary to the roads, and a major component of the economic circulation, was transportation by water. The Great Lakes were becoming a vast corridor to and from the interior of the continent. From the 1820s steam power became common on the lakes, and by the middle of the 1840s there were regular steamboat schedules for many waterfront towns. Toronto, for example, was headquarters for a dozen steamboats, so it had daily connection to Kingston, Hamilton, Niagara, Queenston, Port Hope, Cobourg, and Lewiston and Rochester (in New York State), and as well as other places en route. Hamilton had two steamers daily to Toronto, and another to Queenston and Niagara. Kingston had ten daily steamboat connections. Even centres that were more remote, if on navigable water, had the advantage of long-distance connections. During the shipping season a steamboat plied between Windsor, Goderich, and Owen's Sound (as it was known in 1845, then very much the frontier). Places on Lake Simcoe, whence the mail was still carried by horseback, had steamboat service between Orillia, Barrie, Holland Landing, and some smaller places, during open water.[14]

The shipping fleet did not grow rapidly, although each improvement of the waterways encouraged the building of larger craft. Douglas McCalla has calculated that between 1845 and 1854 something like forty-four steamships, 123 sailing vessels, and forty-one barges can be identified as having been in use in Ontario Great Lakes waters. The mean size of the vessels (well over half of which were not registered in 1849) was about ninety-two tons, with steamships being twice the size of sailing craft, on average. Kingston and Toronto were the major ship-owning centres, and there was a sharp contrast between Lake Ontario, where some ships were over 300 tons, and Lake Huron, where Goderich's fleet of eleven sailing vessels averaged forty-four tons.[15]

Along with the upgrading of shipping went improvements of the waterways. Canals were caught up in the 1820s, as the railways were less than a generation later, in a feverish parade of ambitious proposals. Planning for canals was fuelled by the availability of practical canal-building technology in the 1820s, and by the example of and competition with projects underway in the nearby States. From 1825 the Erie Canal was an effective siphon in and out of New York State from the south side of the Lakes.

Ontario responded first with the building of the Rideau system during the 1820s, purportedly more for military than commercial purposes and largely financed by the British government. The Rideau Canal, completed by 1834, was never commercially important, but it was not alone in this. McCalla has determined that most of the traffic on the Welland Canal – opened in 1830 as Ontario's answer to the successful Erie Canal – "in most years was internal American trade … and it thus served as an adjunct to the Erie Canal. Other than helping to pay operating costs of the [Welland] canal, such traffic had no developmental impact within the province."[16]

The canals on the St Lawrence River were gradually improved and enlarged until, by 1848, there was a "seaway" of nine-foot draft reaching to at least the west end of Lake Erie. Further to McCalla's comment, Thomas F. McIlwraith argues that, in the race to capture the trade of the Great Lakes basin, the government on the British side of the Lakes built canal capacity greatly in excess of what was needed for Canadian registered shipping; and if there was a winner in the race it was not on the British side.[17] What canals did for Ontario, as railways did twenty years later, was to bring in large amounts of capital for construction and many related jobs. Canal construction employed many of the numerous immigrants, and incidentally raised the wages for labour. Because of the ease of attracting capital for major engineering works, both canals and, later, railways were built on a scale not justified by the traffic.

Canals continued to be perennial favourites with promoters. The lower Grand River was canalized in the early 1830s. Although above Brantford the canal provided an outlet only irregularly because of undependable flow in the river, it gave rise to a proposal to link the Grand and Thames rivers to form an "Erie Canal" stretching from the Welland Canal to Lake St Clair (almost duplicating Simcoe's military road of the 1790s).[18] The persistence of canal fever is illustrated by the projected Toronto and Georgian Bay Ship Canal, which was visualized as following a route similar to the ancient Toronto Carrying Place (portage) up the valley of the Humber River. This mid-century proposal was perpetuated in George R. Tremaine's atlas of 1860 and in the York County atlas of 1878 (on the Vaughan Township map) long after the same hinterland was well served by roads and railways. The canal was never built.

The enticement of canal building left little room for concern about the natural environment, but of course, in the immediate area of the construction, the ecological impact was severe. The course of the canal required huge amounts of excavation, and streams and wetlands were commandeered, obliterating habitat for fish and other water-dependent species. In some cases, as with the Rideau and the Welland Canals, major drainage flow was diverted, and new channels cut.

If one visualizes communications as an overlay on the topography of the province – a landscape of communications – the roads and water routes formed a gradually expanding framework along which mail, printed material, and government directives could travel. A final complement to the landscape of communications before the railway, and one that would ultimately challenge it in influence, was the telegraph, which reinforced the routes of major roads and added an electrical jolt to the movement of communications. The telegraph appeared rather suddenly in the 1840s, and more than any other technology (apart, perhaps, from steam power), bridged the prerailway and railway eras. A single telegraph line was extended from Buffalo in 1846, along the north side of Lake Ontario, to reach Montreal by the middle of 1847. As with all of the technological advances that followed, its power lay not only in the amount and novelty of the information but also in the immediacy with which it was made available. Through one fragile line it became possible to communicate in a matter of minutes with distant parts of the earth; very quickly it became *necessary* to take account of the outside world. Peter G. Goheen's analysis of the major newspapers "reveals strikingly the mutual isolation of most of the cities of British North America,"[19] even in 1845, on the eve of the completion of the first telegraph line. By 1847 significant changes in newspapers were occurring. Two Toronto editors were printing telegraphed information on Atlantic shipping from Halifax. What Robert Park has classified as the modern newspaper, in which prolix opinion was replaced by news, was beginning to appear.[20] The telegraph's immediate influence remained focused on the Lake Ontario shore; but by 1853 it had stretched to Windsor and into three short branches (see figure 7.1). The telegraph and the railway quickly developed a symbiosis; the two lines usually paralleled one another, and the dots and dashes of Morse's code became the foundation of railway safety during the 1850s. The full elaboration of the role of the telegraph, however, in railway scheduling and safety, as well as in the transmission of information (such as to newspapers and money markets), lay in the future.

WAITING FOR THE TRAIN

As early as 1832 the House of Assembly in Upper Canada was debating the implications of building a short railway.[21] Through the following decades the journals of the assembly were peppered with proposals for railway lines, and ambitious promoters were scrambling to secure territory that no other line was serving and even attempting to truncate the hinterland of a competing line. At the same time investment capital, ignoring boundary lines and oceans, was seeking opportunities far and wide, and the New World was prime territory.

Was the economic activity of the province able to generate a sufficient value of trade and passenger traffic even when the first rails were being laid at the end of the 1840s? The question needed to be asked, but the promoters were sanguine and keen to proceed: in 1845, the *Toronto Mirror* was convinced that "the Railway would do more ... than all other efforts put together ... It would induce settlement in every part, making the utmost distant wilderness accessible"; and J.M. and E. Trout record the optimism of the well-known firm of Peto, Brassey, Jackson and Betts in 1851 that they could build a railway from Montreal to Kingston "at a rate which would, by their own estimate, produce them the same profit they had made in England and on the Continent of Europe."[22] This was to become the Grand Trunk Railway, and if Messrs Peto et al. realized a profit on their involvement, the public (i.e., the Province of Canada) "never derived and never will derive a single dollar."[23] But the infectious enthusiasm was widespread, and from the late 1830s Ontario was a society waiting for the train.

The public was slow to calculate the cost, bemused as it was by the "wonderful transformation" and "civilizing tendency of the locomotive."[24] The steam engine in its various forms was a source of great power that could be moved to a chosen location. Unlike earlier fixed sources and even the transporting force of the wind at sea, the steam engine was not only mobile but also predictable and controllable. It could be harnessed to move previously unheard-of loads on land, including huge pieces of machinery, and it began to neutralize what had been physical limitations on where industries could be located.

Despite the talk of railways, it was the plank road that characterized the 1840s. It excited road builders (and very quickly the travelling public) as the macadamized road had done in the 1830s. Plank roads seemed particularly suited to Canada, where wood was so plentiful. Unlike most of the railways, plank roads generally were profitable during the first few years of operation, before they began to incur insurmountable maintenance costs. The capital investment was much smaller and the pleasure of driving horses on a new plank road was universally enticing. The difference in investment required is illustrated by comparing the Cobourg Railroad Company, which had a capital of forty thousand pounds (1834), and its successor, the Cobourg and Rice Lake Plank Road and Ferry Company, which had a capital of six thousand pounds to provide the same link through a different medium. The abortive Cobourg Railroad Company had even been the recipient of a loan of ten thousand pounds, said to have been the first instance in Canada of government financial aid to a railway.[25]

Proposals for railways were often intertwined with plans for plank roads, either as a dual system or with the plank road as a backup proposal in case the railway did not attract enough investors to launch it. The Cobourg and Rice Lake Plank Road and Ferry Company is an example of a backup enterprise. Similarly the Toronto and Lake Huron Railroad Company, having failed to build in the five years following its incorporation in 1836, petitioned the legislative assembly in 1841 for an amendment to its act "to confer on the said Company the Right of forming a planked or macadamized Road upon the same conditions, and on the same terms, on which they are now authorized by law to construct a Rail Road ... The object of the Company is ... to open up an immense Country, at present deprived of access to Market, and to secure to *Toronto,* and the other Ports on Lake *Ontario,* a portion of the great and increasing Trade of the *West,* and a participation in the Lake *Huron* Fisheries ... at present monopolized by the people of the *United States.*"[26]

The select committee appointed to consider the petition agreed with the proposal to build a plank road, but recommended "that in stead of running the Road *Northward* to Lake *Huron,* it should take a *Westerly* direction and be terminated on Lake *Huron* about 50 or 60 miles *North* of *Goderich.*"[27] As with the majority of railway proposals, this one failed to produce either a rail or plank road. In 1871 the Trouts characterized these failed dreams to have been "in advance of the times." A revived version of this company, however, was chartered, in a bid to cover all the bases, "to construct a Rail, Planked, Macadamized or Blocked road" in 1845, with the terminus to be at any point on Lake Huron. This charter too expired. A company with a similar name – the Toronto, Sarnia (later Simcoe) and Lake Huron Railway Company, in 1858 named the Northern Railway of Canada – eventually carried out the essential terms of 1836 by 1855, despite the bizarre geographical contortion in its charter "to run from the City of Toronto to some point on the southerly shore of Lake Huron, touching at the town of Barrie on the way."[28]

The London and Gore Rail Road Company, in the southwest part of the province, was assented to by the legislative assembly on 6 March 1834. It was chartered to build "a single or double track, wooden or iron railroad ... and to employ thereon either the force of steam or the *power of animals*, or any mechanical or other power."[29] The line was to link London with Burlington Bay to the east and with the navigable lower Thames River and Lake Huron in the west. It was a time of vigorous experimentation in railways, both technologically, with such things as attempting to use wood for tracks, and financially, in feeling out what kind of hinterland was necessary to maintain the business. By this time there was a burgeoning of lines in the eastern United States, and many entrepreneurs entertained the ambition to be first off the mark in Canada. In addition to the London and Gore proposal, the Cobourg Railroad Company was incorporated in March 1834; the Erie and Ontario Railroad Company in April 1835; the Toronto and Lake Huron Railroad Company, the Huron and Ontario Railroad Company, and the Niagara and Detroit Rivers Railroad Company in April 1836; and the London and Davenport Railroad and Harbour Company in March 1837. From the mid-1840s there was another burst of charters. In 1846 charters were granted to the Montreal and Kingston Railroad, the Wolfe Island, Kingston and Toronto Railway Company, the Peterborough and Port Hope Railway Company, and the Hamilton and Toronto Railway Company. The Woodstock and Lake Erie Railway and Harbour Company, and the Toronto and Goderich Railway Company were chartered in 1848, followed by the Bytown and Prescott Railway Company in 1850. By the 1840s, designing railways was the nineteenth-century version of a search for Eldorado.

All the same, progress toward actual railway building in the 1830s and 1840s appears to have been ill-starred in retrospect. The route was littered with over a dozen false starts, representing some millions of pounds of capitalization, a few bizarre proposals, charters expiring "non-user," and ruined speculators. Even so, the "spirit of enterprise" remained eagerly willing. As the Trouts pointed out, "if anything further were wanting to show that the Canadian people were thoroughly imbued with the spirit of enterprise in those days, it might be found in a petition addressed to the Legislative Assembly of Canada in 1854. The petitioners ... asked incorporation under the name of 'The Northern Pacific Railway Company'; and sought power to construct a railway from Montreal ... to the mouth of the Columbia River."[30] This might be seen both as a response to westward railway initiatives in the United States and an early declaration of the expansionists' plans for Rupert's Land and the Far West.[31]

Once started, railway building became infectious. In the popular fantasy, there would be great economic leaps once the railway was built; that is, the railway was seen as a fairy godmother's wand. As shown by the numerous failures up to the middle of the century, the truth was almost the reverse. A well-developed infrastructure had to be in place, and sufficient ongoing revenue sources had to be available to cover very large operating costs and a huge debt load. A map of where the original railways were located could be seen as a depiction of the main corridors of economic activity around mid-century. Although there was hope that the railway would be an augury of future economic success because of the stimulating effect of the railways on adjacent territory, routes that might have provided a boost for economically struggling districts, such as a Grand Trunk alternative via Bytown and the Rideau, were avoided in favour of a more direct route through the most developed parts of the province.

The capitalization of the railways was in itself a sign of the times. The very large, previous infusions of capital to the province had been public money supplied for military-related matters, such as fortifications and canals, by the imperial government. Such funding had been an expression of the province's colonial status. The financing of the railways drew on private capital (with significant government encouragement and involvement), which, in concert with the arrival of responsible government in 1848, was a declaration of the growing independence of the Canadas. The preparatory work of designing and surveying the route could be done on a relatively modest budget, but actually building the railway called for huge sums. At this point the railway became for the first time a major economic actor, by enticing overseas money into the province. Ontario had achieved a level of development that provided confidence in the success of a railway and of participation in an increasingly international economic arena. In *Planting the Province,* Douglas McCalla recommends that we "see railways less as the cause of than as a part of a larger development process, extending for decades,"[32] in some respects back to the end of the previous century. McCalla also notes that, as big businesses, railways stood alone until well into the second half of the century, and they did not have a clear role in fostering large industries, although a rail connection was usually important.[33]

The prestige and enticement of railways are revealed through the roll call of important individuals who became involved in the initial successful ventures. Prominent politicians were directly involved in negotiations surrounding the Grand Trunk Railway in the early 1850s, and governments at this period were "bending over backwards" to assist in railway construction. Influential businessmen, such as Allan McNab and William McMaster, and even Judge J.B. Robinson, became deeply involved. The premier engineers in the province, including Casimir S. Gzowski, Thomas C. Keefer, Frederick W. Cumberland, and Sandford Fleming, were active in surveying and giving expert advice.[34] British capital was available in large quantities, especially when yoked with British railway construction expertise in such companies as Peto, Brassey, Jackson and Betts.

The tentativeness in the railway experiments is illustrated by the choice of routes. The first railways followed the corridors that had been established by the major roads. As the railways became more influential, and the railway investors more confident, the lines thrust in directions of their own, anticipating the creation of productive hinterlands. Where the railways were successful, roads were adjusted to take advantage of the railway rather than vice versa. Town surveys began to accommodate train tracks, and the railway routes, like the military roads earlier, came to be the arteries of intensification of land use. The height of development occurred along the railway lines and graded away from them toward less continuous land clearing and remnants of the presettlement ecology.

What is generally credited as the greatest stimulus to the building of the first long-distance railways in Ontario was the Guarantee Act of 1849. Its initial raison d'être was to help fund a rail link between Halifax and Quebec City in the face of a refusal by the imperial government to provide financial assistance.[35] For Britain this was a time of shedding colonial responsibilities; but, in the colonies, the drive to tie together the vulnerable parts of British North America – to link the St Lawrence Valley with the Atlantic – for economic security as well as protection was, if anything, in resurgence. The government of the united Canadas decided to provide a financial platform for the building of its major railways – initiated for the line to the east coast, but also available for other lines within the Canadas. The legislation, "An Act to Provide for Affording the Guarantee of the Province to the Bonds of Railway Companies on Certain Conditions, and for Rendering Assistance in the Construction of the *Halifax* and *Quebec* Railway," was given royal assent on the 30th May 1849 (12 Victoria c. 29). It might be argued that the need for this Guarantee Act, together with the long-term indebtedness of Ontario because of it, suggests that Ontario – although willing – was not yet ready to support railways.

Of course, being ready was not always seen by the politicians to be a necessary condition. There was a strong belief that, instead of prosperity bringing the railway, the railway would in its wake bring prosperity, along the lines of another contemporary argument that rain follows the plough. This belief was illustrated, as G.P. de T. Glazebrook points out,[36] by many urban places across the country that tried to attract railway lines with promises and the offering of grants and benefits of various kinds. The various preparations for a railway line – from the surveying, to chopping trees, to earth and gravel moving, to building structures – likely brought the first infusion of hard, imported cash that most localities had seen, at least since military activity and canal building had faded. Among municipalities, competition to be on the line was rampant.[37] The popular understanding of the survival of the fittest was in evidence before Darwin's elaboration of the theory. Urban centres, for their part, were convinced that their place at the crossroads of the continent could only be assured by attracting at least one major railway line.

An indication of how quickly the railways caught on, once they got started in the 1850s, is found in railway company reports. The Great Western Railway in 1855 (the year it reached Toronto) took in £298,419 (ca. $1,200,000) from passengers, £160,939 from freight and livestock, £16,580 from mail and express, and £1,968 from sundries.[38] These figures suggest considerable popularity, especially among passengers, as the Great Western was one of two major railways (with the Ontario, Simcoe, and Huron Railway, later called the Northern) operating in Ontario by this time. Despite the popularity of the train, fares were not particularly cheap. The Great Western, for example, advertised the following fares from Hamilton to Niagara Falls in December 1853: it cost 2s. 6d. (about fifty cents) to Grimsby, 5s. to St Catharines, 6s. 3d. to Niagara Falls, and an additional twenty-five cents (about 1s. 3d.) to continue by omnibus across the bridge to connect with trains on the United States side, a total distance of about forty-five miles (figure 6.3).[39] Train fares a few years later appear to have changed little when in 1857 the London and Port Stanley Railway charged about fifty cents (roughly 2s. 6d.) for the twenty-mile journey between its two termini.[40]

These fares can be compared to wages: although senior officials in government earned one thousand pounds and more per annum, the

GREAT WESTERN RAILWAY.

TIME TABLE.

OPEN FROM

HAMILTON TO NIAGARA FALLS!

ALTERATION OF TIME.

Trains will, on and after Monday, the 12th of December, 1853, until further Notice, run regularly as follows :

GOING WEST.

DISTANCE.		
	NIAGARA FALLS	11.45 A.M.
9¾ Miles.	THOROLD	12.10 P.M.
11¼ "	ST. CATHERINES	12.20 "
22 "	BEAMSVILLE	12.55 "
26¾ "	GRIMSBY	1.05 "
37¼ "	STONEY CREEK	1.40 "
43 "	HAMILTON	2.00 "

GOING EAST.

DISTANCE.		
	HAMILTON	12.25 P.M.
6¼ Miles.	STONEY CREEK	12.45 "
16¾ "	GRIMSBY	1.10 "
21¼ "	BEAMSVILLE	1.25 "
32 "	ST. CATHERINES	2.00 "
34 "	THOROLD	2.10 "
43¼ "	NIAGARA FALLS	2.45 "

The above Trains will connect with Trains to and from

BUFFALO, ROCHESTER, SYRACUSE, UTICA, SCHENECTADY, ALBANY, BOSTON, NEW YORK, AND INTERMEDIATE PLACES.

Omnibuses will be ready to convey Passengers and their Baggage across the Bridge, to and from the above Trains. Fare, 25 Cents, including Bridge Tolls.

The Train from the Falls will run in Connection with the Train arriving at the Suspension Bridge at 10.45 A.M., from New York. Passengers leaving Hamilton at 12.25 P.M., will be in time for the Express Train leaving the Falls at 3.45 P.M., and connecting at Rochester with the Lightning Express Train, reaching New-York

AT 9 A.M. NEXT MORNING.

RATES OF FARE.

FROM HAMILTON	s.	d.
To STONEY CREEK	1	0
" GRIMSBY	2	6
" BEAMSVILLE	3	3
" THOROLD	5	0
" ST. CATHERINES	5	0
" NIAGARA FALLS	6	3

FROM NIAGARA FALLS	s.	d.
To THOROLD	1	6
" ST. CATHERINES	1	[illegible]
" BEAMSVILLE	3	3
" GRIMSBY	4	0
" STONEY CREEK	5	6
" HAMILTON	6	3

Passengers who do not procure Tickets before entering the Cars, will be charged 7½d. Currency, in addition to the above Fares.

Children under 3 years of age, no charge; Between 3 and 12, half price.

Passengers are allowed to carry 100 lbs. of Personal Baggage free of charge; 1s. cy. per 100 lbs. will be charged for any excess above that weight.

THERE WILL BE NO TRAINS ON SUNDAYS.

It is particularly requested, that any incivility or want of attention, on the part of any of the Company's servants, may be immediately communicated to the undersigned.

NO FEES OR GRATUITIES ARE ALLOWED TO BE RECEIVED BY ANY OF THE COMPANY'S SERVANTS.

SMOKING NOT ALLOWED ON THE COMPANY'S PREMISES.

C. J. BRYDGES,
Managing Director.

HAMILTON, December, 1853.

Figure 6.3 Great Western Railway timetable and fare schedule, 1853 (Toronto Reference Library, Baldwin Room)

doorkeeper of the legislative assembly earned only seventy pounds, the sergeant-at-arms earned one hundred pounds, and junior clerks earned one hundred and twenty-five to one hundred and seventy-five pounds. But perhaps closer to ordinary citizens were the messengers at 7s. 6d. per day (the price of a trip from Hamilton to the United States border), the House page at 6s. 3d. per day, and extra clerks in the survey branch of Crown lands, who could take home 7s. 6d. per day.[41]

Despite the cost of train travel, it offered novelty and comfort, and the fares still compared favourably to those of the stagecoaches and steamboats. The Hamilton to Toronto trip by stage was only six and one-half miles farther than Hamilton to Niagara Falls, but in 1851 it cost 12s. 6d., while the steamboat fare was 7s. 6d. The stage fare from London to Port Stanley was 5s. (twice as much as by train), from London to Woodstock (31 miles) 6s. 3d., and from London to Chatham (70 miles) 17s. 6d. Kingston was said to be 177 miles (other sources said 165 miles) from Toronto; the stagecoach fare was two pounds (i.e., 40s.), while the steamboat cost one.[42] No wonder that passengers flocked to the trains, when fares were a little more than half as much; and no wonder that capitalists espoused the relatively permanent railways over the disappointing nine days' wonder of land travel – the plank road.

It is fair to recognize the railway as the herald of a substantially new period in the development of Ontario, but just as it took time and relevant activities to establish conditions on which railways could succeed, so it took time for the influence of the railway to manifest itself widely. The coming of the train did not automatically wipe out economic difficulties. Evidence from the hinterland of the railway stretching north from Toronto in the early 1850s is a case in point:

> The market in the northeastern United States for a wide range of products extended into parts of Ontario by the late 1840s, and into Simcoe County in 1853 with the opening of the Ontario, Simcoe, and Huron Railroad. However, local marketing agencies had to be established. Without these ... there could be no response to market pressures. Most conspicuous was the failure between 1853 and late 1859 of Simcoe County farmers to respond to the American demand for cattle and sheep because there were no marketing agencies in the county.[43]

Far from wiping out economic difficulties, the railways sometimes became part of the problem, counting as they did on large amounts of financial help from towns along their projected routes. The first decade of railway building came to an end in the depressed economic conditions following the Crimean War. What can be said about the influence of the railways is that when railway building finally got under way, settlers in Ontario were given another reason for hastening to invade the last areas of woodland that promised to be usable farmland. During the 1850s pioneers pushed to the limits of the ecumene in the interlake peninsula – and beyond. The hope, stimulated by the 1,400 miles of rails laid in the decade,[44] was that in a few years their crops would be going to market by train, thereby they could overcome another impediment to their farming – distance.

Before the train arrived in the 1850s, connections between urban places were not well developed and an urban hierarchy was barely intimated. After the first two or three major railways were in place, Ontario entered the phase of its most vigorous central place formation and the crystallizing of an urban system.[45]

7 The Urban Role in an Agricultural Colony

Toronto ... may become the thinking head and beating heart of a nation, great, wise, and happy; who knows? ... but at present it is ... a youth aping maturity.

Anna Brownell Jameson, 1838[1]

The town itself [Toronto] is full of life and motion, bustle, business and improvement. The streets are well paved, and lighted with gas, the houses are large and good; the shops excellent. Many of them have a display of goods in their windows, such as may be seen in thriving country towns in England.

Charles Dickens, 1842[2]

For all the supposed progress bestirring Ontario after the War of 1812, there were very few sizable urban places by 1840. Even the ongoing clearing of land by a growing body of farm settlers produced less than three dozen townships with a population density greater than thirty-six (five or six families) per square mile (see figure 3.3C). Incorporation is an indication of the ambitions of an urban place and usually a signal of growth. There were few centres incorporated before 1850, as illustrated by figure 7.1, although there was a small flurry of incorporation in 1850–51 when the first railway routes and telegraph lines were being laid out.

The population in urban centres of one thousand or larger only accounted for 17 per cent of the total population of just under one million in 1852. In the 1848 census, only five centres were distinguished as a town or a city, from the smallest, at the two geographical extremities – London (4,668) and Bytown (6,275) – to larger places on Lake Ontario – Kingston (8,416), Hamilton (9,889), and Toronto (23,503). These five centres accounted for just under 53,000 of the total population of 725,879. Earlier censuses – back as far as the first published returns, from the population count of 1824, which appeared in the *Journals of the Legislative Assembly* in 1828 – had considered it appropriate to recognize only the capital, Toronto (called York until 1834), as a separate urban centre. In 1824 Toronto (York) had just under 1,700

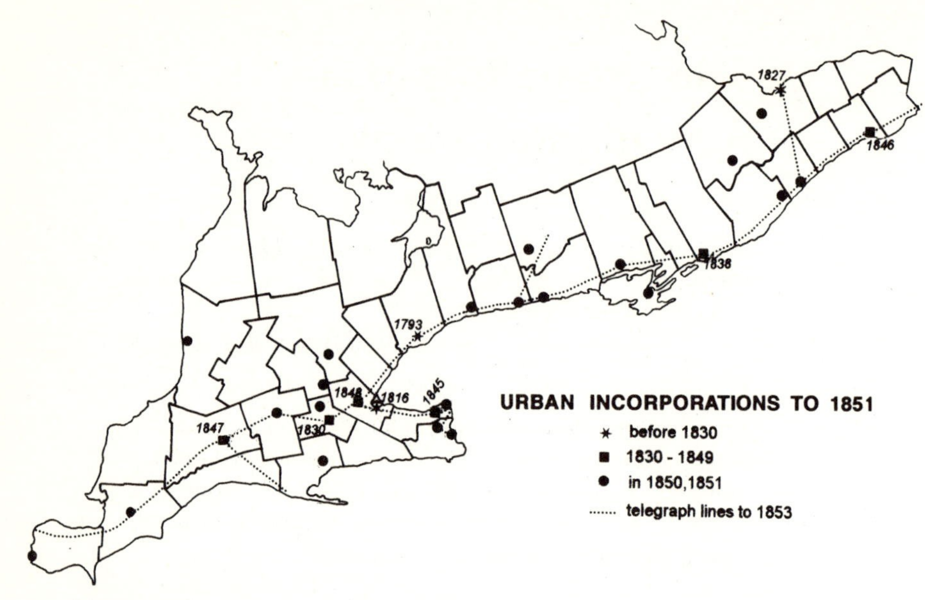

Figure 7.1 Urban incorporations to 1851, and telegraph lines to 1853, with county boundaries (*Economic Atlas of Ontario*, plates 98, 100; telegraph lines from Thompson, *Wiring a Continent*, map following p. 240)

residents, and by the 1825 tally, it had lost a few, at 1,677. Toronto (York) was second in size to the militarily important Kingston, and only surpassed it in the early 1830s.

Clearly some growth was occurring in most of the urban nuclei across the province. Walter Cowan, in 1833, was of the opinion that in the five years since his arrival as an immigrant, York and Dundas had doubled, and Hamilton had tripled in size; and of his local service centre, Galt, he claimed "a few mornings after we came I counted 25 smoking chimneys ... Whether there is 50 now or not I cannot tell, but I believe the buildings are doubled."[3] Of Galt's neighbour, Paris, a dozen miles down the Grand River, it was said in 1832, "there are a saw and grist and wool mills. It is just two years since the first stake was driven in, and now there are about seventy large frame houses and many stores."[4]

If the appearance of a "primate" urban place is a necessary condition for the inauguration of an urban network, then Ontario was struggling toward the first stage of inauguration at the end of the 1830s. Toronto (York) began to grow rapidly in this decade – from a population of 2,800, at the opening, to 9,252, when it declared itself a city in 1834, to 13,092 in 1840. It was only in the 1840s that Toronto began to confirm its primacy and to consistently record a population more than twice that of any other urban centre in the province (what was then Canada West).[5] At the census of 1851–52, Toronto had a population of 30,775; its nearest rivals were Hamilton, at 14,112, and Kingston, at 11,697 (see the rank-size distribution in figure 7.4 and related discussion later in this chapter).

Population comparisons can give further perspective on the relative importance of the larger urban places. Kingston in 1852 was approximately equivalent in population to two well-developed farming townships (which could reach 5,000 to 6,000 in that part of the province), while Hamilton would have numbered close to two or three of its neighbouring townships. Toronto was about equivalent to four townships in its vicinity, some of which contained 7,000 to 10,000 inhabitants. In a meeting based on "Rep by Pop," the towns would not have loomed particularly large. Ontario was overwhelmingly a rural, pioneering agricultural society even into the 1850s, and any urban functions or industrial development owed their existence almost entirely to serving the agricultural economy: the farming environs dominated the village or town rather than the other way around. Even Toronto was somewhat boxed in, in 1851, by woodland (as described earlier), which visitors must have found unseemly for the budding primate city.

With the growth of the villages and towns, the overall population density in many of the townships was rising during the 1840s. Calculations based on the census of 1851–52 reveal 120 townships with a density of thirty-six or more persons per square mile (in contrast to the three dozen such townships of 1840). The major impact on the natural environment at the mid-point of the century was still from the activities of rural settlers, especially in cutting trees and burning the debris. On one occasion Susanna Moodie found herself face to face with the frightening unpredictability of a clearing fire that had run wild near her house, with "red forks of lurid flame as high as the tree tops, igniting the branches of a group of tall pines that had been left standing for saw-logs ... The air was filled with fiery particles which floated even to the door-step."[6]

THE FUNCTIONS OF URBAN PLACES

In the mid 1840s there were a hundred places across Ontario that had some activity to which the label "industrial" might be attached, although in the majority of cases the activity was very rudimentary. In addition there were somewhat footloose activities, such as saw milling, potash making, and shingle making, as well as blacksmithing, fabric making, brewing, and distilling, that were often located outside an urban nucleus. Whether in or out of an urban place, the industries ran the gamut from cloth factory – which probably could be as insignificant as one or two looms, but was often combined with carding and fulling – to foundries, agricultural machinery factories, and coach works. Industrial these enterprises may have been in the strict sense, but large factories were almost unknown. Generally, value-added activities were focused on products demanded by an agricultural population in the early stages of clearing and working land and building the structures of a farm. Value-added activities would have included grist milling, making basic farm and household tools, wagon making, and construction products such as mortar, doors, and sashes. The census of 1851–52 bears testimony to the wide distribution of these activities inside *and* outside named urban places. Even grist milling, which required a major capital investment, was not concentrated in towns and cities to the exclusion of rural locations.[7] Most marketing was very local, and before many years had passed the largest part of the local demand would have been satisfied, leading to an economic reorganization that coincidentally became tied to the developing system of railways.

The distribution of urban centres in 1845–46 (illustrated in figure 7.2) takes the form of an embryonic network. Figure 7.2 is a redrafting of the frontispiece map in Smith's *Canadian Gazetteer* (and should be compared with the foregoing map of the ecumene, figure 6.2). The apparent honesty and thoroughness of Smith's work as a gazetteer leads one to presume that he personally visited most of the districts in the province and made use of the roads indicated on his map. Smith described 262 urban places, 241 of which were on his map (not all appear in figure 7.2). Many were no more than hamlets; here are some examples:

Adelaide

A small village in the Township of Adelaide, situated on the road from London to Port Sarnia, eighteen miles from London. It contains about 120 inhabitants, and an Episcopal Church.
Professions and Trades. – One distillery, two stores, two taverns, one waggon maker, one blacksmith, one shoemaker, one tailor.

Sparta

A Settlement situated near the south-east corner of the township of Yarmouth, six miles east from the plank road. It contains about sixty inhabitants, two stores, one tavern, chair factory, and blacksmith. There is a Quaker meeting house and a Baptist chapel about midway between Sparta and the plank road.

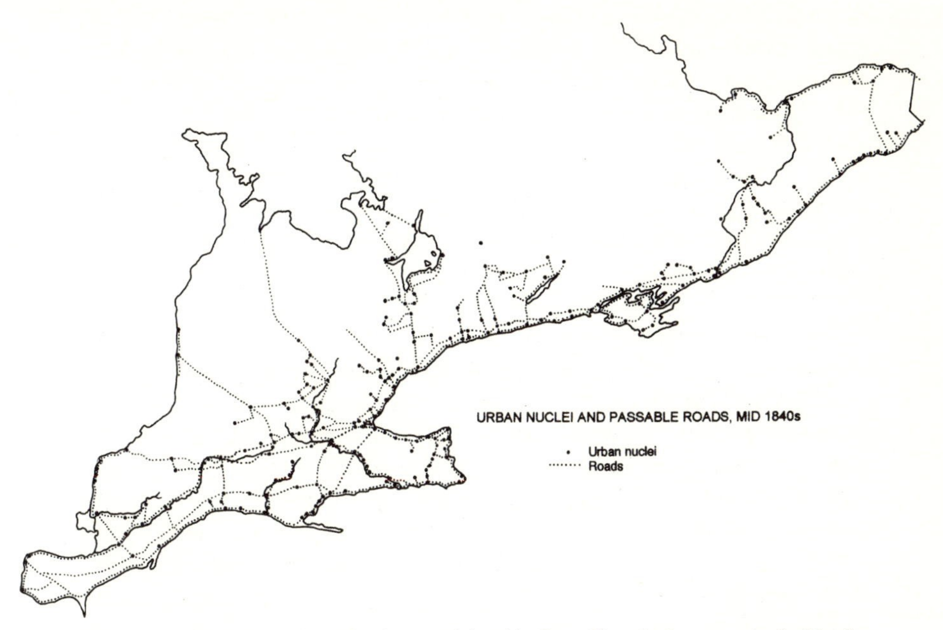

Figure 7.2 Passable roads and urban nuclei, mid-1840s (frontispiece map in Smith, *Canadian Gazetteer*)

Marshville

A small village in the township of Wainfleet, situated on the Grand River feeder of the Welland Canal, ten miles from Port Colborne. It contains about sixty inhabitants, grist mill, two stores, one tavern, one blacksmith.
Post Office, post three times a week.

Stoney Creek

A Village in the township of Saltfleet, pleasantly situated on the road from Hamilton to St Catharines ... There is an Episcopal Church a short distance from the village. Population, about 160.

Jedburgh

A small Settlement in the township of Dumfries, situated on Cedar Creek, a branch of the Nith, about a quarter of a mile from Ayr. It contains about thirty inhabitants, one grist and saw mill, one distillery, one blacksmith.

Brownsville

A small Village in the township of King, one mile and a half north-east of Lloydtown; contains about sixty inhabitants. There are in Brownsville one grist and saw mill, one tavern, one store, one blacksmith.

Warsaw

A Village in the west of the township of Dummer, situated on the Indian River, fifteen miles north-east from Peterborough. It contains about seventy inhabitants, grist and saw mill, carding machine, two stores, and two taverns.

Santa Cruz

A small Settlement in the township of Osnabruck, situated on the St Lawrence, fifteen miles from Cornwall. It contains two churches – Episcopal and Presbyterian; and one tavern.[8]

Some of the hamlets were to grow much larger, and the overall urban proportion of the population would continue to increase until surpassing the rural about the time of the 1911 census. It appears that, in parts of the province, urban places also were beginning to grow at the expense of the rural areas; at least, there were over a dozen rural townships that reached their maximum population by 1851. In fact, there is indirect evidence of the rural-urban shift getting under way by the 1830s: the unexpectedly large number of women in Kingston and Toronto suggests the migration of young women from the country to find employment in the towns (see chapter 3). The provincial population as a whole maintained a majority of men until well after the middle of the century.

While some of the larger central places had a dominant economic specialty, they all also had a greater variety of economic activities than most smaller towns and villages. In 1845 Bytown was "principally supported by the lumber trade"[9] but also had the administrative functions of the district capital, which included the district court and the surrogate court, the sheriff's office, the clerk of peace, the district clerk, the treasurer, the coroner, the collector of timber duties, and a land agent. The population was said to be about 7,000. Apart from five bank agencies and a few dozen commercial establishments, divided between Lower Town and Upper Town, Bytown had two grist mills, two saw mills, three foundries, six tanneries, two breweries, one coach maker and five wagon makers, five blacksmiths, one soap and candle factory, and numerous mobile craftsmen, such as tailors and shoemakers, and small-shop craftsmen, such as tinsmiths and saddlers.

The economy of Kingston in the mid-1840s had reverted to its traditional function as a military and shipping base when its role as capital of the united Canadas had been snatched away. It retained considerable importance as district town, which involved (according to Smith) providing the services of the district court, the sheriff, the clerk of peace, the district clerk, the treasurer, the collector of customs, inspector of licenses, emigration agent, and registrar of the County of

Frontenac. Kingston, like Bytown, was a composite, with a main nucleus close to the harbour and facilities for the shipping industry, and three or four suburbs nearby, one of which was adjacent to Fort Henry and another of which was close to the penitentiary. All told, the population was between 11,000 and 12,000. Within the town scene were two modest developments of mineral wells for medicinal purposes. There were many retail shops, four bank agencies, and ninety-four taverns. In the industrial line there was a shipyard, a steam grist mill, three foundries, three tanneries, four breweries, two coach makers and five wagon makers, seven blacksmiths, one marble factory, and a variety of other activities. There were nineteen steam-driven lake vessels owned at Kingston, as well as many schooners and barges.[10]

Toronto was the "metropolis," with 19,706 people in 1845. It too was a district town with governmental functions such as those found in Bytown and Kingston, but with the addition of the superintendent of schools, probate office, clergy reserves' office, commissariat office, ordnance office, Royal Engineers' office, marriage license office, Indian office, emigrant agent, and board of works. As a further reflection of its importance as a political and cultural focal point, Toronto had ten newspapers, and there were six banks or bank branches, and 107 hotels, inns, and taverns. Its industrial sector seemed somewhat underdeveloped for the size of its population. It housed one steam grist mill, four foundries, three tanneries, thirteen breweries, eleven distilleries, two coach makers and thirteen wagon makers and wheelrights, thirty-seven blacksmiths, one cloth factory, makers of stoves, nails, axes, fanning mills, and pumps, as well as many peripatetic trades. As mentioned earlier, the city had daily steamboat connections with Kingston, Hamilton, Niagara, Queenston, Lewiston, Rochester, Port Hope, Cobourg and intervening places, and regular omnibus connections with the city's hinterland, to the north and the west. These were the "nerve pathways" along which the messages and materials of the nascent urban network were beginning to move.

Hamilton was also a district town and had daily steamboat connections with various places on Lake Ontario, and stagecoach connections in all directions. Its population in 1845 was 6,475. It included among its employed those in various kinds of governmental roles, including the personnel of the district court and surrogate court, the sheriff, the clerk of peace, the treasurer, the warden, the superintendent of schools, the district clerk, the registrar of the County of Wentworth, the inspector of licenses, the collector of customs, Crown lands agent, and coroner. Hamilton had four dozen shops, four bank branches, and sixty-five taverns. A large sailing craft of 330 tons and eleven smaller schooners were owned at Hamilton. Already there was more emphasis on industrial activity than its population might suggest, although Smith mentions neither grist nor saw mills. There were, however, 81,597 barrels of flour exported through the port in 1844, as well as nearly 330,000 board feet of lumber, not to mention 200,000 staves. In the town there were two foundries, two tanneries, three breweries, three coach makers, thirteen blacksmiths, one marble and stone works, two soap and candle factories, and four dozen peripatetic or small-shop craftsmen (although, as noted in chapter 5, these operations did not figure prominently in shipping from the harbour).

London was capital of the London District, in the southwestern lobe of peninsular Ontario. It had planked or macadamized roads connecting to the east, west, and south (Port Stanley), and there were daily stagecoaches on most of those roads. Its population in 1845 was about 3,500. The government offices necessary to the running of a district also were found in London, including those of the district and surrogate courts, the clerk of peace, the sheriff, the treasurer, the district clerk, the deputy clerk of Crown, as well as a coroner, and the Crown lands agent. London had four bank agencies, three dozen retail shops, and eighteen taverns. London's more substantial industrial activities consisted of one grist mill, one saw mill, two foundries, three tanneries, four breweries (one of which is today an international giant), two distilleries, one carding machine and cloth factory, one carriage maker and five wagon makers, fifteen blacksmiths, and a number of the more mobile trades. Many of the townships for which London was the central place had only been opened for farm settlement for twenty to thirty years, and were still deep in the process of farm making.

In the 1840s a conglomeration of only 1,000 people could be described, in the Ontario context, as "a considerable village." That was how immigrant William Smith depicted Paris, on the Grand River. Indeed, Paris, with its excellent water power potential, was a promising place, with two large grist mills, a mill for grinding plaster of Paris (used mainly as a lime field dressing), one saw mill, one foundry and plough factory, one tannery, one woollen factory, one fulling mill, five wagon makers, one brewery and three distilleries, six blacksmiths, and five cooperages, making up the main facets of its industrial activity. This was the result of fifteen years of entrepreneurship since the founding of Paris.

Only two urban places came close to the five largest places in population. St Catharines, with a population the size of London's, had gained greatly through both commerce and water power from its location on the Welland Canal. It had four grist mills, resulting in "an immense quantity of wheat being annually converted into flour," as well as a shipyard, a foundry, a tannery, a brewery and distilleries, and other small factories. It had daily stagecoach connections on the main road from Hamilton to Fort Erie, including daily postal delivery, as well as other less regular services to nearby ports on Lake Ontario and to the Grand River Valley. It had three bank agencies, and was "a place of much trade."

Cobourg, with a population of 3,367 in 1845, was slightly smaller than London. It was the district town for the Newcastle District, east of Toronto, and it had been a centre for the settlement of the central north shore of Lake Ontario, along with Port Hope, for over forty years. It had a well-rounded industrial base, including three grist mills, two saw mills, a large new cloth factory (designed for a work force of 200), three tanneries, a brewery and distilleries, five wagon makers, and a number of other small workshops. That there were four machines for planing lumber – more than Smith showed for any of the other sizable urban places – indicates the upgrading of structures in the region from original log shanties, and perhaps suggests the demand for finished lumber in the cities. There were daily connections by road to the east and the west, and north to Rice Lake. During open water there was a daily steamboat link from Rice Lake to Peterborough, and from the newly refurbished harbour on Lake Ontario to various ports in Ontario and the United States. Cobourg also had agencies for the Commercial Bank and the Bank of Montreal.

Being a district town (or county town after the Baldwin Act of 1849) was obviously one of the most important functions an urban centre could perform. A district town had a standard assemblage of government appointees on government salaries, and there would have been a number of additional businesses or professions to service and complement the government administrators.[11] The choosing of the location for a government town remains something of a mystery. Easy accessibility for a majority of the tributary population should have been an inherent principle in the choice, as it had been for centuries in the identification of parish and county centres in the old country. The choice of location involved some measure of centrality, with practical accessibility being crucial – a requirement also well established in earlier settlements in North America.[12] The prime reason for accessibility was the location of the courthouse and jail: in fact, the site for the district or county town could be predetermined by the existence of a courthouse, which would by its presence foreordain the location for the district offices and other public institutions, such as a registry office, and a house of refuge (or industry). Failing such a pre-emption of the decision, the legislated procedure for identifying a government sub-capital relied on the governor, who was empowered "to issue a Proclamation … naming a place … for a County Town, and erecting …

a Provisional Municipal Council."[13] This wording probably reflects the outline of what was long-standing practice, with the governor having ultimate decision-making authority. The choice of capital at any level must have been the subject of a great deal of lobbying. Where facilities were in a rudimentary condition, a county council could, by a related act, frame a by-law to establish the site of the shire hall, courthouse and jail, and other public facilities (and thus the county capital).[14] A similar process applied at the township level, and once again accessibility and centrality were important considerations. Sometimes the choice was obviously the result of compromise, as in the case of Vaughan Township, where the township hall and offices were located in the geographical centre (at Vellore), far from any of the budding urban centres. In this case the township functions did not lead to significant urban development.

CIRCUMSCRIBED ROLES

The overwhelming impression one gets from the gazetteer's sketches of the urban nuclei in the mid-1840s is that, except for Toronto and a few minor exceptions in the larger port towns, these centres were almost totally engrossed in serving their own circumscribed hinterlands. The localization of economic life is underlined by Lord Durham's astonishment at the "isolation of the different districts from each other."[15] Urban places were receiving, perhaps processing, and exporting the products of the woods – though only Bytown and, to a lesser extent, Belleville serviced the more remote timber shanties – and of the farms, but support was restricted to nearby points of exchange because of the difficulties of travelling. This was the condition Jacob Spelt captured in the 1950s through his application of the "umland" concept to mid-nineteenth century Ontario. An umland was an area of urban influence to which Spelt gave, for purposes of mapping, a radius of 4.5 miles from the urban centre. This figure was significant as the (approximate) maximum distance in most parts of the province around 1840 that a farmer and wagon could travel to deliver a load to town and get back home the same day. The umland, therefore, would be the practical tributary area for an agricultural service centre.[16] Large or small, the urban places of Ontario in the 1840s were serving an agricultural society keenly engaged in creating a farming landscape. This work would not be finished in most areas for the better part of another generation, by which time the clearing of the tree cover would have been completed (and overdone in many older parts), most necessary farm structures and equipment would have been acquired, and the local centres would have lost much of their original raison d'être.

The services and supplies provided by the urban places to their farming umlands led to little in the way of inter-urban connections, except for trade arrangements for exporting local agricultural products and bringing in exotic goods. At the beginning of the 1840s Ontario was in a rudimentary condition, in which the only widespread organizations were fundamentally rural, including the agricultural societies and the Orange Lodges. The temperance societies, which were both rural and urban, and the few mechanics' institutes in towns, were not at this time significant exceptions. Although there were various kinds of government services provided across the province (as indicated in the sketches of the larger urban places, above), most of them were superimposed, and operated with great difficulty. The postal service was probably the most important to farm life, but daily delivery was only available to centres close to the few major roads. After the middle of the decade various influences – the telegraph, the newspapers, the planking and macadamizing of roads, the planning and surveying of railways, the introduction of universal censuses, the development of a new educational bureaucracy – conspired to encourage connections beyond the local umland. By the end of the 1840s a mindset that could contemplate participation in a provincewide, if not international, economy was taking shape.

A GRADUALLY EMERGING URBAN SYSTEM

"Emerging" is a fair general description of Ontario in the 1840s. Urban conditions were clearly undergoing change. Many small places were growing rapidly on the strength of the initial farm making still going on across the province, especially in the new areas to the west and north (a growth of small nuclei that appears to have occurred, perhaps as one might expect, nearly a generation earlier in Quebec).[17] The large places were developing roles in international trade, notably with the United States. At the mid-point of the nineteenth century, a report for the Senate of the United States pointed out that "the trade of the United States with the [British American] colonies has increased to a surprising extent since 1830, and particularly since 1846 ... The exports to Canada alone, in 1850, were equal to the whole amount exported to Sweden, Prussia, Holland, Portugal, and Mexico united."[18] The economic activity in these years was heavily dependent on the booming agriculture and timber harvesting of Ontario. The stimuli and rewards filtered back to the many brokers and forwarders, the merchants, the millers and manufacturers, and eventually to the rural population where most production had begun. The developmental impact can be visualized by comparing the depiction of Ontario's

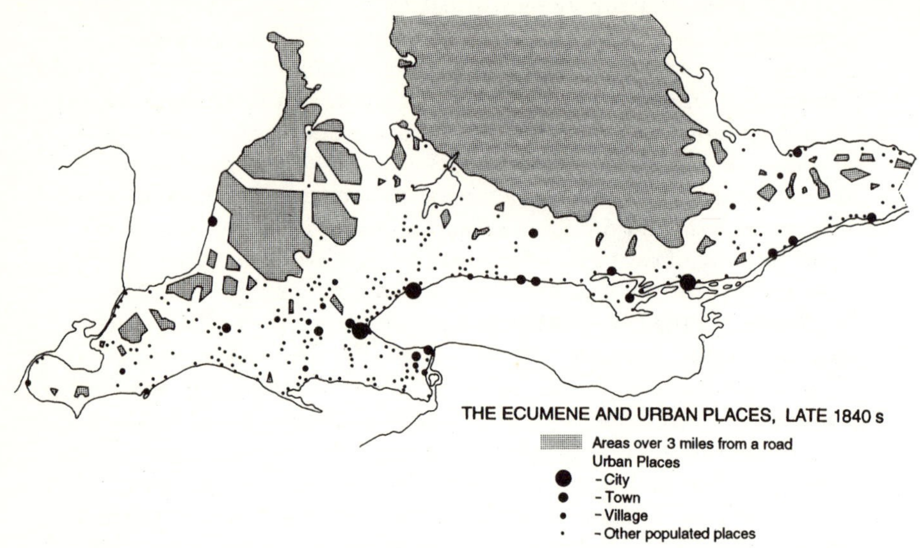

Figure 7.3 Putting the urban infrastructure into the ecumene: the ingredients of settlement expansion (Whebell, "Robert Baldwin," map 1)

urban places and ecumene in the late 1840s (figure 7.3) with the depiction of the mid–1840s shown earlier in this chapter (figure 7.2). The pushing back of the wilderness by 1849 and the emergence of a three-tiered urban hierarchy (city, town, village or hamlet) can be discerned. These were the beginnings of an urban system, just as there had been moves toward other kinds of comprehensive systems – of education, of "modern" municipal government, of communications, and so on, through the spectrum of social infrastructure.[19]

The urban "system" during the last decade of this study is depicted in figure 7.4, in terms of the rank-size rule based on population. Toronto is just emerging as the primate city: in 1852 it is over twice as large as Hamilton, the next largest, whereas in 1842 it had not theoretically "emerged." But the graph also reveals that in the decade of the 1840s the smaller places seem to have been growing slightly faster than the mid-range places, even faster than the large but ill-starred Kingston.[20]

After many false starts and financial losses, the railways were actually on the way by 1851. The Guarantee Act had emboldened many municipalities to become stockholders, and construction had begun or been revived – as with the Great Western Railway, from Niagara to the Detroit River. "Operations ... are now proceeding with energy ... in the confident expectation of completing the road in about two years,"[21] wrote

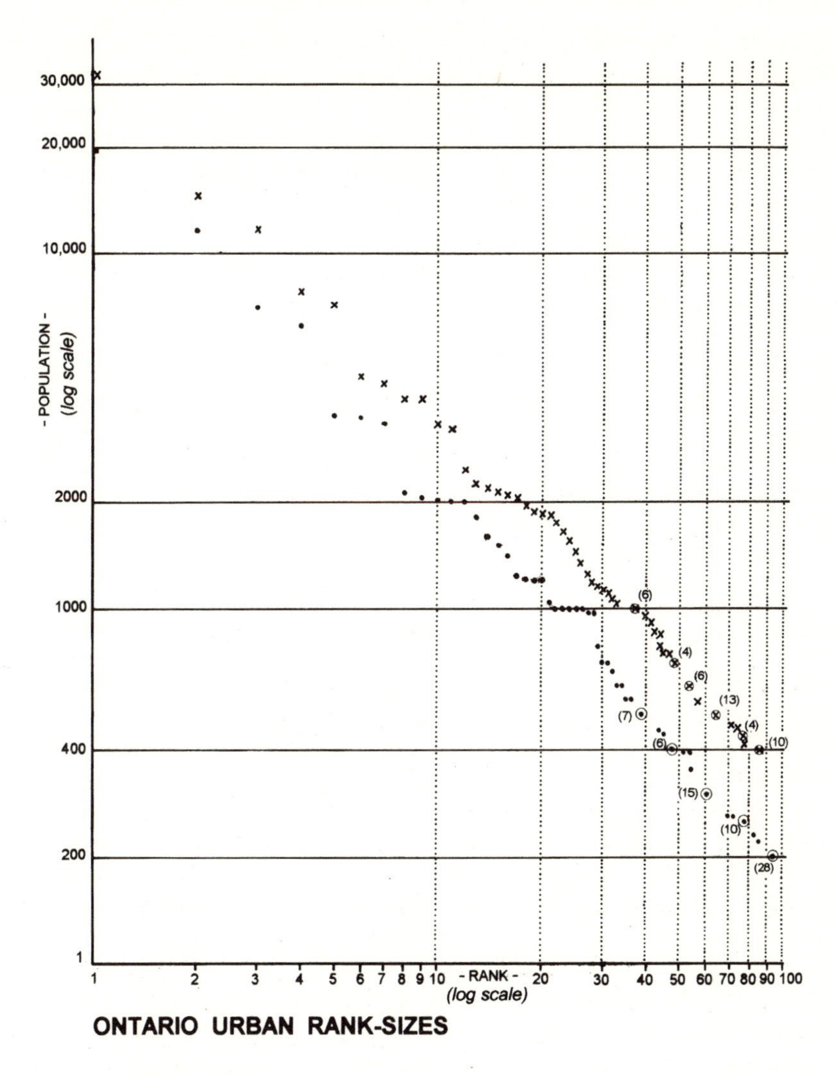

Figure 7.4 Rank-size distribution of urban places, ca 1842 and 1852 (Smith, *Canadian Gazetteer*, for early 1840s; Smith, *Canada*, for places below ca 1,100; Canada, Board of Registration and Statistics, Census of 1851–2, vol. 1, no. 1)

William Smith in 1851. Despite a large measure of optimism and an increase in population, however, people in the towns and villages along the planned railway routes could not be sure of the completion of the railway or of its impact once completed. In later decades, when railway construction was a matter of course and its effects were well known, the railway appeared to cast ahead of it a kind of "aureole of excitement" that led to the establishment of opportunistic functions that would come into their own when the train actually arrived. Around 1850 the urban centres do not seem to have responded in a clear fashion, perhaps because of the many previous false alarms. It might be possible to identify some hints, however, of preparations for the railway.

As the headquarters for the Great Western Railway, Hamilton was assured of a place on the line. Its 1851 population of 14,112 was more than double that of 1845. Hamilton had made some adjustments that would help it take advantage of a rail connection: it had added two banking agencies, for a total of six; it had multiplied the number of commission merchants/forwarders from two to thirteen; it had increased the number of foundries to four (while the number of blacksmiths diminished); and it had added two grist mills, two breweries, two coopers, a saw mill, and a telegraph office – all within six years. No wonder that it was glad to be called an "ambitious little city."[22]

London was to be farther down the line and, although its population had doubled, to 7,035, its expectations of life with the railway were not as clear. It had one more foundry than in 1845, and now had one commission merchant; but, it appears to have lost a brewery, a tannery, and two carriage and wagon makers.[23] These losses may have been a temporary result of devastating fires in the town in the mid-1840s. London was the only one of the five major towns not connected initially to the new telegraph system.

Bytown had a telegraph connection to Montreal, perhaps because of its status as "the head quarters of the lumber trade of Canada."[24] But, Bytown did not experience unqualified growth: according to the census of 1851–52, its population was 7,760, a mere 11 per cent increase over 1845. William Smith said that there was not "the same appearance of busy bustle that prevailed some six years ago."[25] Indeed, there were two fewer bank agencies, and the apparent loss of a grist mill and some wagon makers, blacksmiths, and forwarders. There was a small increase in the number of foundries and breweries, however.

Kingston had experienced a kind of epiphany early in the 1840s when it was chosen as capital for the united Canadas. This led to a rapid increase in size and to frantic activity in building and services. When the capital function was taken away in the middle of the decade, Kingston's fortunes were deflated, never to rebound to the peak of optimism of 1841. The writers of a recent history of the city, in putting the best face on the story of a city lacking in spectacular growth, opt for the essential characteristic of "'continuity' – in particular, continuity of function and continuity of form."[26] In 1851 Kingston seems to have been still struggling: its population, at 11,697, was little different than six years earlier, and it had fewer steamboats registered, fewer coach and wagon makers, blacksmiths, and coopers.[27] Although it was on the telegraph line, Kingston – like Bytown and most of the eastern third of the province – was not participating in the growth that was enlivening the Home District and the head of Lake Ontario.

Toronto was the keystone in the central arch of growth. The tenor of the times ("an age of progress") was encapsulated by the remarks of Governor General Lord Elgin, at the sod-turning for the Ontario, Simcoe and Huron Railway, which was to reinforce the links of the city with its rich hinterland to the north (quoted in chapter 1). At the same ceremony, the mayor of Toronto saw in the hinterland "resources for profitable enterprise, to which it would be difficult ... to assign a limit."[28] Toronto's progress was well enough established by the 1840s that losing the capital function to Kingston seems to have had little effect on the city. By 1851 Toronto was "running with the hounds," confidently awaiting construction of the railway from the east and watching extension of the line to the north. In six years its population had increased by over 50 per cent, to 30,775; indeed, in just over one year it had grown by 20 per cent (from Smith's figure of 25,166 in 1850).

Toronto, like Hamilton and Kingston, had gas lighting, and it also had a water supply pumped from Lake Ontario. It had a number of foundries, factories making such things as stoves, pumps, edge-tools, paper, clothing, and even pianos; and it had breweries and distilleries, some of which were to grow into large, long-lasting businesses. Substantial changes in the fabric of the city were beginning to take place.[29] An image of some sophistication is roused by the listing under "Professional Men, etc," which included Charles and John Ambrose (and a number of others), "professors of music," "Kane, Paul, artist," "Hamel, T., portrait painter," and "Deslandes, Mons and Madame, ladies' seminary."[30] Toronto was well endowed with educational, religious, and medical institutions, including teaching hospitals and a large new lunatic asylum. The elaboration of occupations between 1833 and 1851, when the population grew from approximately 9,000 to 30,000, is illustrated by Peter G. Goheen in his study of Victorian Toronto (see table 7.1).

While the population of the city had increased less than three and a half times, most of the occupational categories had increased much more. The structural expansion of the city is revealed by the more than fivefold increase in the building trades, the more than twelvefold increase in miscellaneous trades, and the number of labourers. There was a sixfold increase of "Professionals," generally the occupations

Table 7.1
Occupations in York/Toronto, 1833 and 1851

Occupation	Number of individuals 1833	1851
Building trades carpenters, masons, joiners, etc.	191	1,007
Merchants and shopkeepers	157	594
Clothing trades tailors, hatters, shoemakers, etc.	129	911
Service trades barbers, blacksmiths, saddlers	117	235
Governmental, judicial, military officers	90	102
Professional clergy, lawyers, doctors, teachers	74	421
Keepers of taverns & lodging houses	70	159
Food trades bakers, butchers, brewers, etc.	60	332
Transportation trades sailors, carters, watermen, etc.	40	385
Metal work trades iron founders, tinsmiths, silversmiths, etc.	26	311
Clerks	21	478
Miscellaneous trades coachmakers, tanners, soap makers, cabinet makers, etc.	47	587[a]
Labourers	56	3,193[b]

Source: Goheen, *Victorian Toronto*, 51
[a] Of this number, 355 were in the woodworking trades
[b] This total includes 352 male servants and 1,264 female servants. (It seems likely that the basis for this category was not the same in both years)

requiring the most training. An indication of the move beyond small individual enterprises toward some concentration of large businesses is seen in the proliferation of clerks.

There is other evidence that indicates even more clearly Toronto's gradual involvement in an urban network. One expediter of this involvement was the telegraph line that linked Toronto to the west through Hamilton and Buffalo (and Windsor by 1853), and to the east through Kingston and Montreal. In Toronto there were twenty-five agencies for fire, life, and marine insurance companies with head offices in Hamilton and Montreal; London, England; Hartford, Conn; New York City, Oswego, Ogdensburgh, and Buffalo, in New York State; and Columbus, Ohio. The city also had eighteen commission merchants and forwarders, not including twenty-eight businesses explicitly engaged in importing.[31]

One might have expected railway fever to have generally gripped the province at the end of the 1840s, engendering contrasts between places that were obviously slated for a train connection, on one hand, and places that had no such expectation, on the other. But the stimulus of the railway was not everywhere apparent. For example, Cornwall was in a location on the lower St Lawrence that would obviously have placed it on the route of the westward advancing Grand Trunk Railway. Although it gained one or two doctors and lawyers, a bank agency, and two grist mills between 1846 and 1851, in other respects not much was happening in Cornwall, whose population had decreased slightly. William Smith described it in 1851 as "a neat, quiet, old-fashioned looking place ... Cornwall is not a place of any great business ... Cornwall remains rather stationary."[32]

A town with which Cornwall might be compared was the far-inland Smith's Falls. Although "a flourishing Village" in 1846, it had no promise of an early railway connection, and the Rideau Canal on which it was situated was not proving to be an effective key to prosperity. Like Cornwall, it did not seem to be making much progress, but it was not losing any more population than its sister town on the St Lawrence corridor.

A somewhat different result seems to have occurred further west. Although Galt, on the Grand River, was not on the proposed routes of either of the main early railways, its population in 1845 had more than doubled by 1850, to 2,200; it had added a couple of grain mills and an axe factory, a second bank agent and a second newspaper, as well as two barristers. Galt secured a branch line to the Great Western Railway by 1854 – over a year before that line reached Toronto. Meanwhile, its older upriver neighbour, Berlin (later Kitchener), had made little advance on the mid-1840s, and had to wait some years for a railway connection.[33]

By 1851 there were just under three dozen urban centres in Ontario with 1,000 or more residents. These centres were either on the major roads, notably John Graves Simcoe's east-west military corridor, or they had good access and shipping facilities on the Lake Ontario-St Lawrence system, or were supply depots for late frontier settlement (namely, Goderich, Barrie, and Peterborough). All these places, Galt among them, were probably scheming to gain a rail connection at an early opportunity, and some would take on excessively heavy financial burdens to attract the train. About a third of the largest places in 1851 turned out to be "persistent gainers" over the next century and a quarter, including Toronto, Hamilton, London, and Bytown, with Kingston just missing this category.[34] Most of the gainers were located in the great corridor that paralleled the north shore of Lake Ontario and continued to Windsor, where Simcoe's main military road ran and the east-west rail lines followed, or they were attached to the other vital transportation route – the Welland Canal.

For long-term growth it was a distinct advantage for Ontario towns to have been relatively large at an early date. Early growth, as argued by D.W. Kirk in 1949, was almost entirely attributable to Simcoe's military plan of the 1790s,[35] the only significant exception in the mid-nineteenth century being Bytown, which was itself the product of a later military plan. The effect the railways were to have on the urban network, as a result of searching out markets and areas of production, was primarily to solidify the embryonic hierarchy that existed in the middle of the nineteenth century.

This was not an evolution without drama: the stakes were high, many prominent actors were involved, there were many sub-plots based on competition among the railways, and almost all centres had variations in their fortunes. The stakes were also high for the woodland environment. A strong relationship developed between the growth of urban places and the clearing of land, which spread in tentacles further and further north and west in the woodland environment of early Ontario. The obverse of these "advances," of course, was the retreat of the original land and wildlife complex.

8 Conclusion: A New Land, Handmade

I consider the Country to be of immense value ... The *immediate advantages* ... to Great Britain are that it is a soil, climate, & situation better adapted to allure American settlers than any she possesses, that the very circumstances of its peninsular situation is likely to prevent emigrations ... till the whole Country shall be filled up; and that this *condensation* of a numerous, industrious Agricultural People is the real force of a nation & the most desirable strength of a frontier Colony ... British Customs, Manners, & Principles in the most trivial as well as serious matters should be promoted and inculcated.

John Graves Simcoe, 1791[1]

In the 1850s Ontario was still a British colony not far removed from what Simcoe had envisioned sixty years earlier. The colony had inherited much of British culture along with tens of thousands of colonists; but rather than a clone of the old country, it had become an amalgam of New World and Old – a compromise between British allegiance and administrative structures on the one hand, and North American know-how and flexibility on the other. By mid century the province was opening for settlement the last of its land that was promising for agriculture, although few previously settled areas could be described as "filled up." There were developments that Simcoe could never have visualized, such as the initiation of the railways, and perhaps he would have been dismayed at the rapidly strengthening interchange with the republican upstart to the south. While faithful to the empire, Ontario was beginning to drift from the centre of the old country orbit, and express its own definition of colony.

THE EXTENT OF TRANSFORMATION

Simcoe's goals had been very much in the tradition of imperialism. He believed passionately in the idea of a mother country with subordinate colonies. The relationship with British North America very quickly became one of economic imperialism (if indeed it did not start as such), although by 1853 the classic mercantilism had faded. In addition to the economic and political expressions of imperialism, there was an ecological imperialism, which is the special interest of this book. This was ecological imperialism in the sense highlighted by Alfred Crosby – the import (and export) of alien plants, diseases, and pests that upset indigenous relationships, often in catastrophic ways[2] – and, more all-encompassing, the ubiquitous damage to and re-creation of the land. By the 1850s the mature woodland that had covered Ontario south of the Canadian Shield was generally over one-third gone, and during the next generation would be nearly three-quarters gone (on the way to an extreme of over 90 per cent by the First World War). In its place grew up a more controlled, regimented scene of superimposed geometry that reached its zenith in the gridironed towns and cities.

With the gradual excision of the trees, the habitat for fauna was profoundly modified. The wild animals – many of which found themselves in conflict with the aims of agriculture, from black bears and deer to the passenger pigeon and the lowly vole – were in retreat. The deadly clashes between humans or their domestic livestock and the wild animals were merely visible extensions of a largely unseen and unstudied drama of ecological transformation that reached down to the microscopic level. Birds and insects, the panoply of ground-dwelling animals, the species in the ground, and aquatic life – all were affected, some extirpated. The closest that the general public came to contemplating this transformation was to wonder if removal of the forest could affect the climate. When the conclusion was in the affirmative, it was often presumed that change would be in the beneficial direction of drier and warmer, although this was wishful thinking rather than a result based on acceptable evidence.

One critical element of the habitat that came under relentless modification was the water on the land. Quite apart from the desirability of good water for humans and their animals, all the wildlife depended on water. Whole natural communities were built around available water, whether on the surface or in the ground, but the aim of agriculture was to drain off "excess" water so that the land could be used for growing crops. This process continued steadily through the stages of taming the land, until by the twentieth century the counties of southern Ontario varied in the amount of their original wetland drained from at least 20 per cent up to 80 per cent. Drainage legislation beginning in 1868 simply increased the pace of a process that went hand in hand with land clearing.

It appears that the many hard-working and honourable pioneers sketched in this book overran their corners of the eastern woodland with little evidenced sensitivity or sentimentality. In comparison to modern practices, settling Ontario was a matter of clear-cutting the region, spread over a generation or two, with the kinds of damaging side effects documented in forestry today, such as loss of soil and damage to organisms of all sizes. But it is not right to attribute this damage just to the farming pioneers; because agriculture and, in a different way, timber exploitation were the foundations of the whole society and its economy, people in businesses in town as well as those on the land were part of the process of transformation. Of course, the settlers did not view their efforts as a kind of "scorched earth" policy. They worked within the mentality of "progress": what they did was accepted as proper, and the society at the time gave them credit for it.

An attempt is made in figure 8.1 to provide a summary picture of the essence of the transformation by displaying the proportion of the land in each township that was under some kind of agricultural use (i.e., well cleared of original vegetation, except for stumps in newer areas), and the number of industries recorded in the 1851–52 census for each township. The cleared land was described in the census as "under culture," which included farm crops, improved pasture, "gardens," and probably fallow land that had been seeded in the past.[3] The industries could include operations with as few as one worker; even for activities that required a sizable capital investment, such as grist and saw mills, the range could be great in the number of millstones or saws, and thus in the number of employees. All the counties, except Bruce and Middlesex, reported additional industries not attributed to a particular township (thus not included in figure 8.1), such as shingle, chair, lath, sash, and carriage factories, oatmeal and plaster mills, brickyards, potteries, and asheries. The data are for the townships (i.e., primarily rural areas), and figure 8.1 does not attempt to depict the industrial activities in the towns (see chapter 6). The incipient railway lines have been added to figure 8.1, the intention being to indicate the range of pressures on the natural environment from three major activities: land clearing, industrial demands on local resources, and the stimulus to change provided by the new transportation medium.[4]

Figure 8.1 exhibits the outcome in 1852 of the demographic upsurge of the preceding decade. The general pattern of growth was already intimated by the densities of 1840 (figure 3.3C). In figure 8.1 the northwest end of Lake Ontario reveals the incipient Golden Horseshoe, at this time based on strong agricultural productivity – especially in wheat – on land rapidly approaching complete clearing. Land clearing has also been vigorously pursued in the Grand River Valley, just west of the head of Lake Ontario, and industrial establishments in this valley and along the north shore of Lake Ontario have burgeoned. There is a focusing of railway lines on this economically active part of the province.

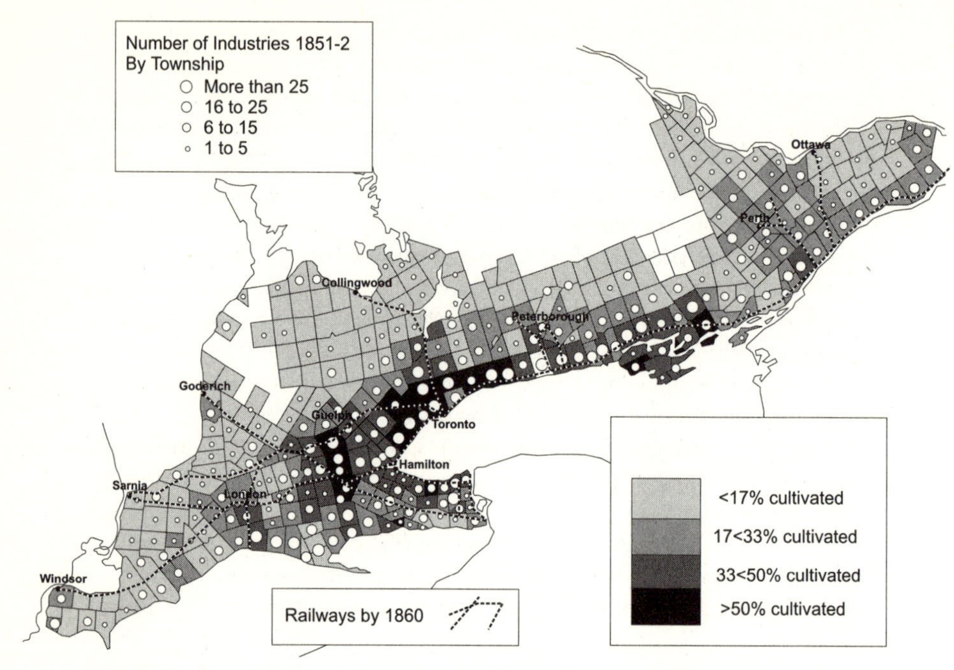

Figure 8.1 Extent of transformation, ca 1852 (Canada, Board of Registration and Statistics, *Census of 1851–2*, vol. 2, no. 7; Kerr, *Historical Atlas of Canada*, 48; Trout and Trout, *Railways*)

The oldest Loyalist areas, in Niagara and around the east end of Lake Ontario and the upper St Lawrence Valley, still demonstrate the advantage of having been the first areas developed. Although the level of cultivation is lower than at the northwest end of Lake Ontario and in the Grand River Valley, with many townships showing less than one-third of their land cultivated because of a lack of good farmland, the number of industries is generally large. The district with the oldest Euroamerican settlement, in the far southwest corner of the province, reveals a low level of cultivation and only a modest industrial development.

The area southwest of London appears generally backward, with even the valley of the river Thames displaying only a slightly larger number of industries than in newly opened townships. The arch of townships north of Long Point on Lake Erie is at an early stage in preparing land for cultivation, but the prevalence of industries (and the level of affluence shown on figure 5.7) hint at major gainful employment, in this case related to the extraction of timber for sale in the United States.

Apart from the newly settled fringes, the other region of relatively low development is the far eastern interfluve of four dozen townships between the Ottawa and St Lawrence Rivers. This backward sector could be deemed aberrant in light of its early settlement, the townships on the St Lawrence, around Perth, and on the lower Ottawa having had at least one full generation on the land. Even the line of the Rideau Canal, from Ottawa to Kingston on the east end of Lake Ontario, does not show marked advancement, although many townships have basic farm service or timber-related industries. The low proportions of cultivated land betray the exceedingly variable soil quality and, between Perth and the head of the St Lawrence, the starkly negative effect of the Canadian Shield and its broad limestone "apron."

Except for the almost immutable presence of the Shield and its limestone borders, the Niagara Escarpment, and the Great Lakes shorelines, the economic and demographic landscapes of Ontario had been nearly reversed in the seventy years since first Euroamerican settlement – from initial high points on the St Lawrence and Bay of Quinte, the Niagara frontier, and the Detroit River to the new heights of the 1850s in the central and northwestern littorals of Lake Ontario, and in the Grand River Valley.

By 1852 Ontario had a population of one million and a leading commercial centre of over thirty thousand (Toronto). The sod-turning for the first railway emanating from Toronto occurred in October 1851, and goods were being railed to and from Lake Simcoe townships north of the city by 1853. The province had been linked instantaneously to the outside world since 1847 by the telegraph line from Buffalo to Toronto and Montreal. The roads had been improving, thanks to various technical innovations, since the late 1830s, so much so that William Smith could travel from end to end of the province in 1850 preparing his second socio-economic survey, and Baron de Rottenburg could plan and carry out a mapping of the roads from the Ottawa River to Lake Huron.

Ontario had been developing, though hesitantly, from the foundations laid down by an early plan – Simcoe's long-forgotten military plan of 1792. Part of the reason for Simcoe's persistent influence was his training as a military engineer, for which a geographical sensibility was required. Simcoe began contemplating the organization of the colony before he left London in 1791, but it was because he traversed and assessed the province in the field shortly after his arrival that the plan he devised for Upper Canada fitted the landscape so well. Through his strategy, based on military roads and the regional centres they linked, Simcoe gave certain naturally endowed parts of the province a head start. Once launched, these centres and roads went on to incise their influence on the practical organization and operation of the province, so that the main features of Simcoe's plan were still dominant in 1851 (and indeed in 1951).[5]

Surprisingly the railways did not have a substantial impact on the general geographical relationships that had crystallized over the previous half century. As C.F.J. Whebell and others have argued, the railways by and large reinforced the corridors of development that had arisen from Simcoe's roads and nodes.[6] A comparison of the map of the first railways (in figure 8.1) with the redrawing of William Smith's map of the roads and urban places in 1846 (figure 7.2) provides visual underpinning: the railways reinforced the major east-west orientation, paralleling the Lake Ontario shore and extending on to London and Windsor, and Simcoe's south-north strategic creation – Yonge Street.

The railways did participate, however, in the engineering of a new scale of economic organization. The first half century or so of Euroamerican settlement had been preoccupied with building an agricultural landscape. Economic activities had been focused on clearing away the trees and then creating the fundamental structures of the farms within relatively small, somewhat isolated, areas. Most such areas spawned a small service centre to satisfy the needs of establishing the farms from the ground up.[7] By the 1850s, most farms in the province were well established and, except for blacksmithing, grist milling, and perhaps carding and fulling, the role of many of the smallest service centres began to wane, simply from having served their purpose. Mechanization, including the steam power that freed many activities from needing a water power site, began to appear at an increasing rate in the decade before the railways (see the distribution of steam-powered mills on figure 5.7). Mobile mechanization and railways teamed up to hasten the demise of small centres by making the towns on the corridors much more efficient producers of goods (including products of traditional workshops) and services. The small centres suffered further eclipse later because they could not compete with the rail-side factories in terms of speed and cost, and sometimes in quality. Even for agricultural producers the railways brought advantages by reducing the effects of seasonality and by making markets much more accessible and varied, and good profits more regular.

The mechanics of transforming the land of Ontario ranged from the axe to the locomotive. It has been argued that over the span of years from 1780 to the 1850s the axe had wrought the greatest change, but by the end of the period under study here, the railway was showing its potential as the major force for change. The population of Ontario looked to the railway for economic betterment, but this would come at a price to the natural environment: trees were cut down in surveying the line; uneven land surfaces were excavated or filled in making the roadbed; fires were started by stray embers from locomotive fireboxes; and the noise and speed of the trains frightened the indigenous wildlife into retreating toward the margins. Compared to the prerailway period, the latter part of the century was to hold more of the same, but with escalating technical power.

THE ETHICS OF MAKING A NEW LAND

Would Simcoe have been bemused or uncomfortable with the way his former colony was evolving in 1853? For one thing, it was steadily moving away from the stripped-down, class-based imitation of the old country that he had visualized. There was no established church, and the affluent upper class that he had counted on to provide leadership and culture had been somewhat discredited by their inability to understand and satisfy the legitimate expectations of the majority of settlers to gain affluence for themselves. But Simcoe's Upper Canada had developed true to form as a settler colony sustained by an environmental ethic in which rampant exploitation was unquestioned.

The clearing away of the woodland was going on apace: the last great timber frontier in Ontario south of the Shield, around the shores of Georgian Bay and Lake Huron, was about to be exploited, and clearing for farms had reached the Shield in the centre and east. Farm settlers almost everywhere had achieved at least basic shelter for themselves and their animals; they were gradually adding to their cropland and fencing their fields; wet areas were being drained; vistas were being opened from one farm to another; and beyond the farm, community structures were being built and roads improved. Robert Gourlay's description of settlers "imprisoned in the woods" was a thing of the past everywhere except the frontiers.

By the middle of the century Ontario was a major agricultural producer. From time to time it had been of importance in the British market, but now, as indicated by J.D. Andrews's report to Congress, there was a wide-ranging trading relationship with the United States, focused especially on agricultural products. This relationship was soon to be made more formal (and more political) in the Reciprocity Treaty. Agriculture was the raison d'être of Ontario, both for the farm settlers and the colonial authorities. Indeed, the farm scene was still in the making in most parts of the province. At the average rate of clearing land, it would have taken most of a generation (about thirty to forty years) to achieve a commercially viable farm of fifty to sixty acres of cropland; thus most of the farm population was busy on the land and in activities related to their locality. Although farmers were represented in the legislative assembly, the governing of the province for most of the existence of Upper Canada seems to have been dominated by administrators and lawyers only indirectly attached to rural life. Lord Durham and his advisers, at the end of the 1830s, caught a glimpse of

> this body of men [that] possessed almost all the highest public offices ... wielded all the powers of government ... disposed of the large number of petty posts which are in the patronage of the Government all over the Province ... The bench, the magistracy, the high offices of the Episcopal Church, and a great part of the legal profession, are filled by the adherents of this party ... they are all-powerful in the chartered banks, and, till lately, shared among themselves almost exclusively all offices of trust and profit.[8]

After Durham there was a broadening of political influence that continued through the 1840s with the rise of non-establishment figures such as Egerton Ryerson, Francis Hincks and George Brown, and the lapsed establishment figure, reformer Robert Baldwin. The old guard had been somewhat discredited, but even members of the new guard were not scions of the dominant agricultural community, although they generally represented its interests more faithfully.

Ontario had elevated its status on the world scene, between 1780 and 1853, from an emergency shelter for refugees to a choice land of opportunity. The largest part of its population, at least beginning with the Simcoe administration in the 1790s, had come willingly and independently. The province was seen as an agricultural frontier, and it displayed many of the characteristics of frontiers in the New World: a relatively young population with high birth and low death rates, a great deal of transiency, many rags-to-riches stories, epic scenes of humans vs. nature, and revolutionary changes over a short period from wilderness to an agriculturally productive scene. But it must not be forgotten that within that apparently positive panorama, as on all frontiers, there was a considerable percentage of hopefuls who did not achieve their dreams: there were diseases such as typhoid and cholera and the endemic fever-and-ague, there was crime, there was religious and ethnic hostility, and there was a rebellion. Although Ontario had a net gain in the immigration-emigration calculus in virtually all the years in the first half of the nineteenth century, in some years in the second half of the century the province was to lose more people to emigration than it gained from immigration.

In the 1840s, especially in the southern townships that had been carrying on farming for forty or more years, the farmers of Ontario were beginning to adapt their practices to the natural qualities of their land and climate. Returning to figure 8.1 (and comparing it with the agricultural regions in figure 5.4), it is evident that this adaptation was creating different agricultural personalities in different areas and, at least in the relatively advanced areas, giving rise to regional distinctiveness. These regions persisted for many decades, some even to the present day, as in the dairy belts of former London District and the upper St Lawrence, the specialty crop regions in Niagara and the far southwest corner, and the largely beef-based mixed farming north and northwest of the Oak Ridges Moraine.[9] Marked prosperity, based on successful agriculture or industry or both, had developed in places along the Lake Ontario shore, and in the middle Grand River Valley. On the other hand, neither agriculture nor local industry were thriving in certain other regions, notably the Ottawa Valley, the Shield, the "northern slopes" morainic lands, and most of the southwest corner except for areas close to the towns.

It is worthwhile to contemplate and if possible learn from the past. The initial period of settlement of the province was a time of heroics and, in retrospect, massive change, but the vaunted "progress" of the assault on the eastern woodland haunts us today. The complete domination of the environment, from chopping trees through to the cavalier use of heavy earth-moving machinery, from the purposeful introduction of alien plants and other organisms to the application of herbicides and pesticides, and the welcoming of human innovations in preference to the natural – all these activities have left a degraded nature *and* a degraded human habitat. Over two centuries we have lost massive amounts of fertile topsoil, we have experienced an increasing number of damaging floods, we have lost many smaller streams and rural aquifers. Species of fish have been wiped out, many urban drinking water sources are under the deadly threat of intractable toxic pollutants, and many of our machines foul the air with respiratory irritants. Perhaps what we can learn from a review of the beginnings of the Euroamerican occupation of this part of the world is that, although we recognize and honour the heroism and determination of the pioneers in their struggle against the old forest, the tough mentality cultivated by them in the opening phase of transformation is no longer appropriate. Today we are at the opposite end of the process of change, when what is needed is reflection and renewal of nature and people. The battle mentality of the early days has become deeply ingrained, a cultural construct expressed in the values of our society; indeed, the natural environment remains something to be overcome in favour of development, the late-twentieth century version of progress. The hope that a lesson can be learned from the past lies in the greatest strength of humans – the ability to reason and foresee, and thereby to change even subconscious mores.

Notes

PREFACE

1 Ontario was officially named Upper Canada from 1791 to 1840, then Canada West with the union of 1841. Apparently because of confusion and resistance to giving up the former names, the Interpretation Act of 1849 made it legal to continue using Upper and Lower Canada in lieu of Canada West and Canada East (see Banks, "Upper and Lower Canada"). The official name "Ontario" came only with confederation, in 1867, but because of familiarity will be the name generally used herein; "province" refers to the territory that became Ontario.
2 Lower, *Great Britain's Woodyard,* 38–41.
3 For a recent Canadian apologia, see Courville, *Introduction,* where full recognition is given to the French, British, and American forerunners, and to the practice of historical geography.
4 Griffiths and Robin, *Ecology and Empire.*

CHAPTER ONE

1 Lord Elgin, at the sod-turning of the Ontario, Simcoe and Huron Railway, *Globe,* 16 Oct. 1851.
2 Craig, *Lord Durham's Report,* 93. For the ideology of progress, see Fallis, Jr, "Idea of Progress"; also Inge, *Idea of Progress.* For progress and its lexicon, see Becker, *Heavenly City.*
3 Read and Stagg, *Rebellion of 1837.*
4 *Historical Atlas of Canada,* vol. 2, plate 14.

5 Gourlay, *Statistical Account*, 299–301.
6 Noel, *Patrons, Clients, Brokers.*
7 Johnson, *Becoming Prominent*, 10–19, ch. 5. For additional detail, see Saunders, "What was the Family Compact?"
8 Whebell, "Upper Canada District Councils Act." The districts were replaced by a larger number of counties a few years later, under the terms of the Municipal Corporations Act of 1849, Upper Canada, commonly called the Baldwin Act (12 Vic., c. 81); see Spragge, "Districts"; and for the evolution of municipal legislation, Rogers, *Law.*
9 Cox et al., *This Well-Wooded Land*, 65.
10 Calculated from Canada, Board of Registration and Statistics, *Census of 1851–2*, vol. 1, table 1; vol. 2, table 7. See tables 6.1 and 6.2 in McCalla, *Planting the Province* for related data, from 1805 and after, based on districts.
11 Quoted in Cox et al., *This Well-Wooded Land*, 51.
12 Canada, *Journals of the Legislative Assembly (hereafter JLA)*, 1846, appendix F, No. 17.
13 Gilmour, *Spatial Evolution of Manufacturing.*
14 Canada, Board of Registration and Statistics, *Census of 1851–2*, vol. 1, introduction; also see *Historical Atlas of Canada*, vol. 2, plate 14.
15 Cf., Spelt, *Urban Development*, ch. 3. For a discussion of the theory of central place hierarchies, with examples from Ontario in 1961, see Marshall, *Location of Service Towns.*
16 Smith, *Canada*, 1:283. Anna Brownell Jameson used almost identical words fifteen years earlier.
17 Rorke, *Forty Years.*
18 U.P. Hedrick, quoted in Cox et al., *This Well-Wooded Land*, 52.
19 Quoted in Jasen, *Wild Things*, 26. The whole introductory chapter is relevant.
20 Bade, *John Muir*, 1:142.
21 Rorke, *Forty Years*, 80.
22 Evernden, *Social Construction.*
23 Jasen, *Wild Things*, 22–6.
24 Even the justification for Canada's first national park in 1887, fifteen years after the first in the United States, was that it would make money: Canada, House of Commons, *Debates*, 3 May 1887, 233.
25 Cf., eighteenth- and nineteenth-century terms: Becker, *Heavenly City*; also Baker and Biger, *Ideology and Landscape*; Dickens and Fontana, *Postmodernism and Social Inquiry.*
26 John Howison ridicules the Scottish Highlanders in Glengarry County and in the Talbot settlement in *Sketches* letter 2, Letter II.
27 *La Rochefoucauld-Liancourt's Travels*, 19; Goldie, *Diary*, 24–31.
28 Smith, *Canadian Gazetteer*, 238ff.

CHAPTER TWO

1 Sauer, "Agency of Man," 49.

2 Glacken, *Traces on the Rhodian Shore,* frontispiece; also see Mc Neill, *Global Condition,* esp. part 2, "Human Condition: an Ecological and Historical View."

3 Kelly, "Transfer of British Ideas."

4 Crosby, *Ecological Imperialism.*

5 In Houston and Smyth, *Irish Emigration,* 158.

6 James Sharp, letter of 2 June 1821, in the Sharp family papers, near Glenmorris, Ont. For imported plants, see Crosby, "British Empire."

7 Coal, imported from Pennsylvania, could only be afforded by industries that produced a high-value product, such as smelted iron. See McCalla, *Planting the Province,* 110. Importation of coal only became practical for the Lake Ontario shore after the opening of the Welland Canal in 1830.

8 Williams, *Americans and Their Forests,* 81.

9 Cronon, *Changes in the Land,* 120–1.

10 For an excellent related discussion, see chapter 4 in McCalla, *Planting the Province,* esp. 60–6; also, Lower, *North American Assault.*

11 Lambert and Pross, *Renewing Nature's Wealth,* 44–7. This history calls the period up to 1841 "The Age of Waste." There are a number of relevant insights in Wynn, "Society and Environment."

12 Archives of Ontario, RG 1, A-I-6, vol. 8, no. 4.

13 Williams, in *Americans and Their Forests,* 144–5, shows that the debate had certainly started by the beginning of the century.

14 Kelly, "Damaged and Efficient Landscapes"; Wood, "Moraine and Metropolis."

15 Snell, *Wetland Distribution.*

16 Smith, *Canadian Gazetteer,* 237.

17 Rorke, *Forty Years,* 24; see also 70–7. A wide-ranging review of this subject is found in Cook, *1492 and All That,* esp. 20–2.

18 Smith, *Canadian Gazetteer,* 238–9.

19 Wicksteed, *Provincial Statutes.*

20 Quoted in Wilkinson, *Lions in the Way,* 59.

21 Radcliff, *Authentic Letters,* 163–6, 168–9, 170–3, 157–8.

22 *Economic Atlas of Ontario,* plate 99. For an authoritative description of Ontario survey practice, see Weaver, "Ontario Surveys." For a diagram of a surveyed township, see figure 5.2.

23 Wycoff, *Developer's Frontier.*

24 Whebell, "Corridors"; Kirk, "Southwestern Ontario."

25 Wood, "Reflections."

26 Copies of both maps can be found in the Toronto Reference Library.

27 Ontario, Department of Planning and Development, *Humber Valley Watershed Report*, 245. The Oak Ridges Moraine is a huge glacial deposit, ca 120 km long by 15 km wide, stretching from near Orangeville east to Rice Lake.
28 Ontario, "Report of the Fruit Growers' Association," 237.
29 Ontario, Department of Agriculture. *Report on the Reforestation of Waste Lands.*
30 Guelph Conference, *Conservation.*

CHAPTER THREE

1 Godley, "High Church," 144.
2 Henripin, *Fertility in Canada*, table B.6, 366, and discussion in chapter 2.
3 Cowgill, "Transition Theory"; Gagan, *Hopeful Travellers*, 71–8. R.A. Easterlin claims to have found a fertility model in which the highest rate of reproduction occurred in new settlement areas at a time of ready availability of land, but "somewhat behind the frontier": "Population Change," esp. 61, 72–3. See also McInnis, "Childbearing and Land Availability."
4 Wood, "Population of Ontario."
5 Bylund, "Theoretical Considerations."
6 The township population data are taken from municipal records (RG 21) in the Archives of Ontario (hereafter AO). Each township clerk was required to report the population in the township by households, naming the head and at least the total number of males and females above and below age sixteen in the household.
7 The size of the family seemed to have little influence on the decision to move; early evidence is in Curti, *American Community*, 68. Hal S. Barron found that in a much older settlement, between 1860 and 1880, households with four or more children were the *most* likely to leave: *Those Who Stayed Behind*, 89.
8 Cf., Gagan's analysis of "the persistent minority" in later Peel County, in *Hopeful Travellers*, 116ff.; also, H.J. Mays's thorough study of one township in "Place to Stand." Allan Bogue reports a remarkably similar finding, what he calls "the persistence of a substantial nucleus of settlers from the first census," in all the townships he studied in the midwest United States, and he points to the inaccuracy of F.J. Turner's notion that "primitive" settlers were replaced by incursions of more "advanced" ones; see Bogue, *From Prairie to Corn Belt*, 26–7. And see a similar qualification of Turner in Mitchell, *Commercialism and Frontier*, 133. Writing on Ontario, Robert L. Jones appears to perpetuate the Turner notion, in *History of Agriculture*, 53–4.
9 This is related to Wilbur Zelinsky's "Doctrine of First Effective Settlement" as applied to national cultural characteristics, in *Cultural Geography*, 13, 20–3.

10 Bouchard, "Family Reproduction." Related discussion will be found in Wood, "Population Change."
11 Gourlay, *Statistical Account*, 22.
12 When these wetlands were later drained, thanks to provincial legislation beginning in the 1860s, they became among the choicest croplands in Canada according to the federal government's Agro-Climatic Resource Index: see McCuaig and Manning, *Agricultural Land-Use Change*, 26, 32–5.
13 For an illustrative analysis of the expansion and retreat of settlement around the Canadian Shield edge, see Osborne, "Frontier Settlement."
14 Cf., Spelt, *Urban Development*, chap. 3 and figs. 6 and 7. Thirty per square mile is a density similar to today's rates in southern Bruce County and in eastern Ontario away from the cities and major highways.
15 For a detailed discussion of this subject, see Wood, "Population Change."
16 Hart, "Spread of the Frontier," 73–4.
17 Clarke, Taylor, and Wightman, "Areal Patterns."
18 Hart, "Spread of the Frontier," 73.
19 Cf. Spelt, *Urban Development*, figs. 6, 7.
20 Cf., McInnis, "Childbearing and Land Availability"; Gagan, *Hopeful Travellers*, esp. ch. 3.
21 In AO, RG 21.
22 See Darrell Norris's illustration of this for two families in "Household and Transiency," 412–14. For evidence of repetitive demographic similarities in new townships on another agricultural frontier, see Bogue, *From Prairie to Corn Belt*, ch. 1.
23 Because the population enumeration was an annual spring obligation of the township clerk, the 1838 count would have been taken three or four months after the main uprising in December 1837.
24 The aggregate figures for 1842, 1848, 1851–52 are found in Canada, *Census of 1870–71*, vol. 4.
25 This census is best referred to as the census of 1851–52 because it appears that the enumeration was actually taken early in 1852.
26 Brunger, "Geographical Propinquity."
27 McBeath's career was reconstructed from unpublished family papers.
28 Benson migration reconstructed from Katz, *People of Hamilton*, 94–111; and Houston and Smyth, *Irish Emigration*, 152–5.
29 For a later version of this exodus, see Brookes and Wilson, "'Working Away'."
30 For the raw data see Martin, *History of the Colonies*, 189, 191.
31 *Historical Atlas of Canada*, vol. 1, plate 68; for native occupance, 1820, see vol. 1, plate 69.
32 Ontario, Legislative Assembly, *Appendix to the Report of the Ontario Bureau*, 25.
33 Howison, *Sketches*, 18–21.

34 Howison, *Sketches*, 167–76, leading to his disquisition on "the natural depravity of man." For a balanced assessment of the distribution of Scottish-born persons in nineteenth-century Ontario, see Clarke and MacLeod, "Concentration of Scots."
35 Akenson, *Irish in Ontario*; Houston and Smyth, *Irish Emigration*, esp. figure 7.5; Wood, "Settlers and the Land."
36 For data on ethnic origin and religion, see Canada, *Census of 1870–71*, vol. 1; for Orange Halls, see [Simcoe County], *Illustrated Historical Atlas*.
37 Donald H. Akenson explores many facets of this puzzle in "Whatever Happened to the Irish?."
38 Armstrong, "Toronto Directories."
39 Hill, "Negroes in Toronto," 74.
40 Nitkin, "Negro Colonization."
41 Law, "'Self-Reliance,'" 108.
42 Traill, *Backwoods*, 74–5.
43 *Historical Atlas of Canada*, vol. 2, plates 32, 33.
44 Smith, *Canadian Gazetteer*, 68–9; *idem*, *Canada*, 2:250–1.

CHAPTER FOUR

1 From a letter to Sir Joseph Banks, when Simcoe was still in London, 8 Jan. 1791; in Cruikshank, *Simcoe Correspondence*, 1:18.
2 Cruikshank, *Simcoe Correspondence*, 1:27.
3 Leyburn, *Frontier Folkways*.
4 Hay, "Property;" Flaherty, *Essays*, esp. those by Banks, Wylie, and Baker.
5 Compare the situation in Ontario with that of southeastern Pennsylvania a century earlier, where the able elite were prospering from and living on their land. Lemon, *Best Poor Man's Country*, 9–13.
6 As indicated earlier, most aspects of a settler's life were governed by the relatively democratic apparatus of the township meeting, at least until the 1840s.
7 McIlwraith, "British North America."
8 Park and Wood, "Poor Relief."
9 Gourlay, *Statistical Account*, 300.
10 Ibid., 240.
11 Ibid., 290–1.
12 Ibid., 284.
13 Milani, *Robert Gourlay*.
14 National Archives of Canada (hereafter NA), RG 5, B 21, vol. 1, parts 1 and 2.
15 NA, RG 5, B 21, vol. 1, parts 1 and 2.
16 Cameron, "Canada Company."
17 NA, RG 5, B 3, vol. 14, fo 76; Ruttan, "Disquisition." For a scholarly reflection on the broad political landscape and reform initiatives, see Whebell, "Robert Baldwin."

18 Canada, Board of Registration and Statistics, *Census of 1851–2*, 1:xxiii–xxxix. The "First Report of the Secretary" is very extensive and not "misleading," as claimed in an interesting article by Inwood and Sullivan, "Comparative Perspectives," 90. For comparison, see [Proudfoot], "Proudfoot Papers," 141. For a view of the economic condition of individuals in early Ontario, see Brian Osborne's assessment of probate records in "Wills and Inventories."
19 Wallerstein, *Modern World-System*, 3:228–30.
20 Goheen, "Changing Bias."
21 Cross, "Necessary Failure."
22 Zeller, *Inventing Canada*, 4, 51.
23 Kelly, "Farm Produce Marketing Agencies."
24 Goheen, "Canadian Communications," 38–9.
25 Canada, *Journals of the Legislative Assembly* (hereafter *JLA*), 1846, appendix F, no. 17.
26 Ontario Association of Agricultural Societies, *Story*, 20.
27 Matthews, "Local Government"; Osborne, "Trading on a Frontier."
28 Jones, *History of Agriculture*, ch. 10; Talman, "Agricultural Societies."
29 From the *Mechanics' Magazine*, in Eadie, "Napanee Mechanics' Institutes," 210.
30 Careless, *Union of the Canadas*, 104–8.
31 Eadie, "Napanee Mechanics' Institutes," 219–20; Klotz, *Review*; Vernon, "Adult Education," 12.
32 Houston and Smyth, "Orange Order."
33 Ibid., 262–3.
34 Gourlay, *Statistical Account*; Smith, *Canada*, vol. 1: business directory.
35 Godfrey, *Medicine for Ontario*, ch. 1, esp. 16–18.
36 Mackay, *Canada Directory*, 58, 363, 373; Smith, *Canada*, vol. 1: business directory.
37 Gidney, "Elementary Education."
38 Curtis, *Building the Educational State*, chs. 2–3. In fact education in Ontario was unusual in its rapid development of a bureaucracy and reliance on experts. By contrast, the politicians competed with the experts in the Canada Board of Works in the 1840s, and balked at reliance on qualified inspectors for oversight of the railway operations: see Baskerville, "Transportation." T.C. Keefer's embittered career as a chief railway engineer, despite his eminence as a publicist, may have been a real-life expression of the anti-expert biases: Nelles, "Introduction," part 5, in Keefer, *Philosophy of Railroads*.
39 Smith, *Canada*, vols. 1, 2: statistical tables for districts.
40 Canada, Patent Office, *Patents*, vol. 1.
41 Sokoloff, "Invention," 374.
42 Fingard, "Poor in Winter."

43 Splane, *Social Welfare in Ontario*, ch. 3.
44 Ibid., 76–7.
45 Ibid., 74. For a further discussion of houses of refuge, see Park and Wood, "Poor Relief." See Canada, JLA, 1849, appendix V.V.V., for a report of £500 given in support of the Toronto House of Industry.
46 NA, RG 5, B 21, vol. 1, pts 1, 2: Hawke, "Answers."
47 The teachers' stipends are from Curtis, *Building the Educational State*, 65, 92.
48 Arnold, "Geography of Colonial Unrest," 135.
49 Graff, "Literacy," 260–1.
50 Prentice, *School Promoters*, 16. See the more recent Houston and Prentice, *Schooling and Scholars*, 31, 95–6.
51 Smaller, "Teachers and Schools"; also Houston and Prentice, *Schooling and Scholars*, Part One.
52 Gidney, "Elementary Education," 3–4.
53 Thomson, "Tradesman's Travels," 140–1. See related reflections in Garland and Talman, "Pioneer Drinking Habits."
54 *Canada in 1832, 1833, and 1834*, 25; also see Canada, JLA, 1849, appendix Z.Z.Z.
55 These three foregoing quotes are taken from manuscripts in NA, RG 5, B 21, vol. 1.
56 Thomson, "Tradesman's Travels," 140–1.
57 Barron, "Genesis of Temperance." For a more recent Canada-wide survey, see Noel, *Canada Dry*.
58 Bonnycastle, *Canadas*, 1:128.
59 Based on analysis of the records of the criminal assizes for 1838, 1842, and 1846 (in Archives of Ontario, RG 22, ser. 134, vols. 7, 8, 9) and district population figures from Canada, *Census of 1870–71*, vol. 4. This discussion deals with charges brought in the criminal assizes (Queen's Bench) – that is, "true bills," having enough substance to be tried before a grand jury. This is not the same as convictions, which represent only a portion of the charges tried.
60 Cross, "'Dark druidical groves,'" 301ff.
61 Hay, "Property," 17–63.
62 Smith, *Canada*, vol. 2: business directory.

CHAPTER FIVE

1 Canada, Board of Registration and Statistics, *Census of 1851–2*, 1: x.
2 Ibid., 1: xxxii.
3 Gagan, "Enumerator's Instructions." The classic work on agriculture in Ontario is Jones, *History of Agriculture*.

4 Wood, "Woodland-Oak Plains."
5 Flader, *Great Lakes Forest*, esp. essays by Bourdo, Ahlgren, Twining, and Williams.
6 Peter Russell found a range of clearing per adult male per year from less than one-quarter to over five acres average by township, throuth the assessments of 142 townships. See "Forest into Farmland."
7 Primack "Land Clearing," 484.
8 Kelly, "Wheat Farming," 103.
9 From a private letter of 1832 in the Cowan family papers, near Galt (Cambridge), Ontario.
10 Ball, "Technology of Settlement," 131.
11 Primack, "Land Clearing," 485–6.
12 Largely from C.H. Danhof, *Enterprise in American Agriculture*, quoted in Primack, "Land Clearing," 485–6.
13 Bade, *John Muir*, 1: 142.
14 Smith diary, Archives of Ontario (hereafter AO).
15 Freure diary, Baldwin Room, Toronto Reference Library.
16 Crawford diary, AO.
17 Clark diary, National Archives (hereafter NA), MG 24, I 149, vol. 1. I am grateful to Prof. Brian Osborne for alerting me to this source.
18 Cf., *Historical Atlas of Canada*, vol. 2: plate 14.
19 In contrast to the urbane engineer Thomas Keefer (quoted by McCalla, *Planting the Province* 132–3), the snow season was anticipated by farmers and the rural population in general as a time of freedom of movement and settling of accounts. See the farm-year diagrams (figure 5.1) for winter visits to mills and market.
20 Walsh, "Consumer Behavior," 224–6.
21 Courville, "Space, Territory, and Culture."
22 Schama, *Landscape and Memory*, 15.
23 Wilson, *Clergy Reserves*; Gates, *Land Policies* chs. 15, 17; Wood, "Reflections."
24 For evidence on one aspect, see Widdis, "Speculation and the Surveyor."
25 Owning land in other districts was a long-standing and widespread custom: see especially figure 2 in Wood, "Reflections."
26 It is not always possible to determine from the Abstract Index the precise disposition of a property. The Essa Township figure is a best approximation, based primarily on the index. The fees applied to bona fide settlers in the early 1820s worked out to approximately 2s. per acre, which was still markedly lower than the standard $1.25 per acre for government land in the adjacent United States. See Gates, *Land Policies*, 123–5, 153–5; [Simcoe County], *Illustrated Historical Atlas of Simcoe*, 28.
27 Consult McCalla, *Planting the Province*, tables 4.3 and 12.2 (appendix 3).
28 McIlwraith, "British North America," 242, and fig. 10.13.

29 Canada, *Census of 1870–71*, vol. 4, table 5, 168.
30 Cf., Canada, *Census of 1851–2*, 2: 42.
31 Marr, “Tenant vs. Owner,” esp. 52.
32 Jones, *History of Agriculture*, 55–6.
33 Canada, Board of Registration and Statistics, *Census of 1851–2*, 1: xxix. According to the agricultural census, the largest county productions were from York and Oxford: Canada, *Census of 1870–71*, vol. 4, table 7, 195.
34 The data for figure 5.4 are printed in Canada, *Census of 1870–71*, vol. 4, table 8, 170. Acreages are based on average yields of crops in the 1840s.
35 Ontario, Agricultural Commission, *Report of the Commissioners*, 2: 523.
36 Alan Brunger and John Clarke have identified differences across Ontario based primarily on demographic features such as place of birth. They have shown that in certain areas there were concentrations of immigrants from a common source, a result of the effort to settle near relatives or friends. Parts of Ontario were identified with particular ethnic groups for generations. Brunger, “Distribution of English”; *idem*, “Distribution of Scots”; Clarke and Buffone, “Social Regions.”
37 *Toronto Mirror*, 17 July 1846, 2.
38 Canada, *Census of 1870–71*, 4:168–9.
39 *Toronto Mirror*, 24 July 1846, 3.
40 Anderson, “‘Irrepressible Stampede.’”
41 Lower, *North American Assault*, diagram 5. Collections of data and various analyses appear in McCalla, *Planting the Province*, appendix B, and in McInnis, “Perspectives,” but the gaps are still significant enough to leave a question about the agriculture-timber shares of Ontario exports.
42 McCalla provides an enlightening analysis, arising from extensive calculations on the available data, in which he demonstrates that, as far as the Ontario economy was concerned, the internal timber market for lumber, fuel wood, and ashes was very significant, even reaching a level comparable to the export of timber down the St Lawrence in some years. See his *Planting the Province*, ch. 4 and appendix B.
43 Head, “Forest Exploitation.” Squared timber was primarily white pine and oak that was exported as large logs. Saw logging used various kinds of trees that were sawed into lumber before sale or export. For an effective illustration of the changes in the timber industry through the nineteenth century, see plate 11 (Graeme Wynn) and plate 38 (Grant Head) in *Historical Atlas of Canada*, vol. 2.
44 Grant Head reinforces the dichotomy between squared and sawn timber production in a recent paper: “Pine Lumber Industry.”
45 Loucks, “New Light”; Wood, “Stage is Set.”
46 McIlwraith, “Rural Roads.”
47 Ball, “Technology of Settlement.”
48 Lower, *Settlement*, esp. chs. 3, 4.

49 Anderson, "'Irrepressible Stampede'."
50 See, for example, Smith's *Canadian Gazetteer*: Cobourg, Port Stanley, Windsor.
51 Anderson, "'Irrepressible Stampede'"; Hudson, "Migration."
52 For saw mills, Smith, *Canadian Gazetteer*, 199; *idem*, *Canada* 1: 288. For percentage of township in woodland, Ontario, Department of Planning and Development, *Humber Valley Watershed Report*, 245.
53 Spelt, *Urban Development*, 73–9.
54 For examples of the evolution of relatively complex contemporary rural industries, see Hardy, *La siderurgie*. On the shift toward more complex industries from the middle of the nineteenth century, see Winder, "Before the Corporation."
55 From a comparison of Smith's *Canadian Gazetteer* and his *Canada*.
56 This and the following examples are taken from Mackay, *Canada Directory*: Woodstock, 471; Brantford, 37; Kingston, 127; Port Hope, 279; Prescott, 284; Cobourg, 64–5; Brockville, 40–1; Ayr, 21; Bytown, 45; Hamilton, 100, 102.
57 The examples in this paragraph are also from Mackay's *Canada Directory*.
58 McIlwraith, "Great Lakes Grain Trade," fig. 9–2.
59 Ibid., fig. 1-2.
60 [Haldimand and Norfolk Counties], *Illustrated Historical Atlas* 61; and McCalla, *Planting the Province*, 108–9.
61 McCalla, *Planting the Province*, 109–10.
62 Newell, *Technology on the Frontier*, 90–7.
63 Smith, *Canadian Gazetteer*, 49. The discussion in this paragraph draws on various parts of Smith's gazetteer.
64 Gilmour, *Spatial Evolution of Manufacturing*.
65 Noel, *Patrons, Clients, Brokers*.
66 Errington, *Wives and Mothers*, 331n12.
67 *Toronto Mirror*, 1 Jan. 1841, 2.
68 *Examiner*, 27 Jan. 1841, 1.
69 Canada, *Journals of the Legislative Assembly*, 1849, appendix V.V.V.
70 NA, RG 5, B 21, vol. 1, part 2: report by the President of Northumberland District Agricultural Society.
71 Smith, *Canadian Gazetteer*, 245.
72 Ibid., 245.
73 See for comparison the value of net farm output in figure 2 of McInnis, "Perspectives," 57.

CHAPTER SIX

1 National Archives (hereafter NA), RG 5, B 21, vol. 1: Treadwell, "Reply."
2 Jameson, *Winter Studies*, 34–5.
3 Pattison, *Historical Chronology*, 4 (my thanks to Dr R. Summerby-Murray for

this reference). See Thomas F. McIlwraith's case study of York County and his argument about the perceived need for good roads in "Adequacy of Rural Roads."

4 Glazebrook, *History of Transportation*, 1:110–13; Pattison, *Historical Chronology*, 229.
5 Glazebrook, *History of Transportation*, 1:115.
6 Traill, *Backwoods*, 31.
7 Glazebrook, *History of Transportation*, 1:115.
8 Cox et al., *This Well-Wooded Land*, 69, 124–5.
9 NA, RG 11, vol. 81, no. 2: Baird, "Specifications."
10 Ellis, *Life on the Old Plank Road*, 9–11.
11 Talman, "Travel in Ontario."
12 NA, National Map Collection. Full bibliographic notation is found in Joan Winearls, *Mapping Upper Canada*, 58.There is a description and colour reproduction of parts of this map in Gentilcore and Head, *Ontario's History in Maps*, 157–9.
13 Smith, *Canadian Gazetteer*, 100. Other towns are listed in alphabetical order.
14 Ibid., 9, 194; the schedules in this paragraph are from the same source.
15 McCalla, *Planting the Province*, table 7.1, table 7.2, 283.
16 Ibid., 124.
17 Glazebrook, *History of Transportation*, 1:79–84; McIlwraith, "Freight Capacity."
18 St-Denis, "Erie Canal"; Wood, "Arks of Absalom Shade."
19 Goheen, "Changing Bias," 190.
20 Ibid., 182, 185. See also Goheen, "Canadian Communications."
21 Glazebrook, *History of Transportation*, 1: 143. George Stephenson's "Rocket" had made its historic trip in 1830, but had been preceded in 1825 by his "Locomotion": McQuire, *Visions of Modernity*, 183.
22 *Toronto Mirror*, 6 June 1845; Trout and Trout, Railways, 68.
23 Trout and Trout, *Railways*, 61.
24 Keefer, *Philosophy of Railroads*, 9,10.
25 Trout and Trout, *Railways*, 54.
26 Canada, *Journals of the Legislative Assembly*, 1841, 616.
27 Ibid.
28 Trout and Trout, *Railways*, 54, 106.
29 Information on proposals in this paragraph is from Trout and Trout, *Railways*, 87 (quote), 54–6, 68.
30 Ibid., 56.
31 Owram *Promise of Eden*, 32–9.
32 McCalla, *Planting the Province*, 215.
33 McCalla, "Railways," esp. 208.
34 Trout and Trout, *Railways*, 57ff., 87ff., 110; Careless, *Toronto to 1918*, 76–81.
35 Glazebrook, *History of Transportation*, 1: 167.

36 Ibid., 167–8.
37 A good example is described in Ennals, "Cobourg and Port Hope."
38 Trout and Trout, *Railways*, 99.
39 From copy of poster in the Baldwin Room, Toronto Reference Library (see figure 6.3).
40 Calculated from the *Report of the London and Port Stanley Railway*.
41 Mackay, *Canada Directory*, 545–8.
42 Mackay, *Canada Directory*, 94, 119, 136, 410.
43 Kelly "Agricultural Geography," 3.
44 Calculated from Trout and Trout, *Railways*, 35–6.
45 Marshall and Smith, "Dynamics of Growth."

CHAPTER SEVEN

1 Jameson, *Winter Studies*, 52.
2 Dickens, *American Notes*, 248.
3 Cowan, Walter, letter of 16 Sept. 1833, Cowan family papers, near Galt (Cambridge), Ont.
4 A traveller, quoted in Smith, *Forks of the Grand*, 22.
5 Canada, *Census of 1870–71*, vol. 4 (review of historical records); re: primacy, see Marshall, "Umland-based Model."
6 Moodie, *Roughing it in the Bush*, 333.
7 Canada, Board of Registration and Statistics, *Census of 1851–2*, vol. 2, table 7.
8 Smith, *Canadian Gazetteer*, 286, 175, 112, 88, 22, 205, 164. For the construction traditions out of which these villages were created, see McIlwraith, *Looking for Old Ontario*.
9 Smith, *Canadian Gazetteer*, 24. The following thumbnail sketches of the towns are based on Smith, *Canadian Gazetteer*: Bytown, 24–5; Kingston, 91–3; Toronto, 194–5; Hamilton, 75–7; London, 100–1; Paris, 142; St Catharines, 177–8; and Cobourg, 34–5.
10 For a full historical context for Kingston, see Osborne and Swainson, *Kingston*.
11 Cf., the Business Directories in Smith, *Canada*, vols. 1,2.
12 Lemon, *Best Poor Man's Country*, 132–5. For the locations of the district and county capitals in Ontario, see map 2 in Whebell, "Robert Baldwin."
13 Canada, 12 Victoria, c. 78.
14 Canada, 12 Victoria, c. 81. A slightly later, related kind of negotiation is covered in Knight, *Choosing Canada's Capital*.
15 Quoted in Spelt, *Urban Development*, 100.
16 Spelt, *Urban Development*, 99, fig. 8.
17 Courville, "Villages and Agriculture."
18 U. S. Congress, *Report on Trade*, 44.
19 Splane, *Social Welfare in Ontario*; Houston and Prentice, *Schooling and Scholars*;

Curtis, *Building the Educational State*; Whebell, "Upper Canada District Councils Act"; Ball, *Building Canada*; Goheen, "Canadian Communications."

20 To secure populations for places not separately enumerated in the census, two useful statistical accounts of William Smith were used. The population cut-off for the early date was 200; for the later, 400. It is possible that the early population figures estimated by Smith (the majority) were truer for 1845 than 1842. Many of the lower population figures for 1852 also were estimates. For both the earlier and later data, where the same rounded figure is attributed to a number of places, that number appears in parentheses in figure 7.4, and the relevant point is circled on the graph. Caution is in order, however, because Smith's gazetteer rounded the populations of small, unenumerated places at the earlier date; the "hard data" for larger places appear to have come from the census of 1842. The figures for smaller places not singled out in the 1851–52 census (generally below ca. 1,100) also were rounded in Smith's new account of that date (*Canada*). Prof. John U. Marshall comments on this graph that Toronto's primacy is "emerging" but is not yet "well-established" and that the faster growth of small places seems to have continued through the 1860s (personal communication). See Marshall and Smith, "Dynamics"; Marshall, "Umland-Based Model"; and Marshall, *Structure of Urban Systems*, 355.

21 Smith, *Canada*, 1: 222.

22 Smith, *Canada*, 1: business directory, 26–32; quote in Katz, *People of Hamilton*, 1.

23 Smith, *Canada*, 1:91, and business Directory, 5–9.

24 Ibid., 2: 357.

25 Ibid., 2: 360.

26 Osborne and Swainson, *Kingston*, 1.

27 Smith, *Canada*, 2: 278–84.

28 *Globe*, 16 Oct. 1851.

29 Gad, and Holdsworth, "Streetscape and Society."

30 Smith, *Canada*, 1: business directory, 61–7.

31 Ibid., 1: 45–60.

32 Ibid., 2: 387. For 1846, Smith, *Canadian Gazetteer*, 173.

33 Smith, *Canadian Gazetteer*, 62, 15; and Smith, *Canada*, 1: business directory, 35–45.

34 Marshall and Smith, "Dynamics," fig. 1, 30–2; also Smith, *Aspects of Growth*.

35 Kirk, "Southwestern Ontario"; further analysis of Simcoe's planning is found in Wood, "Population of Ontario," 94–6.

CHAPTER EIGHT

1 Letter of 30 June 1791, in Cruikshank, *Simcoe Correspondence*, 1: 27–8.

2 Crosby, *Ecological Imperialism*.

3 See introduction to Canada, Board of Registration and Statistics, *Census of 1851–2*. One cannot ignore the reputation, even within the Board of Registration and Statistics itself, of the 1851–52 census. In the words of the secretary of the board, referring to the Ontario returns, "many of the Enumerators proved themselves wholly unfit ... and the negligence and ignorance displayed in the work of these has added materially to the labours of the Office." (*Census 1851–2*, 1: iv.) Nevertheless, this first near-complete official census still seems worth cautious use.
4 The impact of the development of land for agriculture can be visualized through a series of maps showing improved acreages, farm output, and related characteristics, in McInnis, "Ontario Agriculture."
5 Kirk, "Southwestern Ontario."
6 Whebell, "Corridors."
7 Spelt, *Urban Development*, chs. 2, 3.
8 Craig, *Lord Durham's Report*, 79.
9 For a current regional depiction, see Warkentin, *Canada*, 292–317.

Bibliography

PRIMARY SOURCES

Archival Sources

ARCHIVES OF ONTARIO (AO)

Diary of Benjamin Crawford

Diary of Benjamin Smith

Field notebook of surveyor Adrian Marlett (1816–17)

RG 1: Ontario. Crown Lands Papers

RG 21: Municipal Records

RG 22, ser. 134, 135: Ontario. Criminal Assizes Records for 1838, 1842, and 1846

NATIONAL ARCHIVES OF CANADA (NA)

MG 24, I149: John C. Clark. "Diary of the Wind and the Weather." Vol. 1

National Map Collection: Baron Maj. De Rottenburg. "Map of the Principal Communications of Canada West." [1850?] Copy in Robarts Library, University of Toronto

RG5, B3, vol. 14, fo 76: Sheriff Henry Ruttan. "A Disquisition on the Newcastle District." 1840

RG5, B21, vol.1: Sheriff Charles P. Treadwell. "Reply to Governor General's Questionnaire." 18 December 1840

RG5, B21, vol. 1, pts 1, 2: A.B. Hawke, "Answers for the Information of Emigrants ... by the Chief Emigrant Agent." 9 November, 1840

RG5, B21, vol. 1, pt 2: "Report to the Governor General by the president of

Northumberland District Agricultural Society"
RG 5, B26: "Upper Canada: Returns of Population and Assessment"
RG11, vol. 81, no.2: Canada. Board of Works. "Specifications for Earth Work of Plank Road from Lake Ontario to Rice Lake Under the Variety of Situations." Report no. 60, by Nichol Hugh Baird. 2 June 1842

SIMCOE COUNTY ARCHIVES, MIDHURST, ONTARIO
Abstract Index to Deeds

TORONTO REFERENCE LIBRARY, BALDWIN ROOM
Browne, J.D. "Map of Township of York, in the County of York, Upper Canada." 1851
Cane, James. "Toronto in the Province of 'Canada.'" 1842
Diary of Benjamin Freure

Newspapers

Examiner (Toronto)
Globe (Toronto)
Toronto Mirror
Upper Canada Gazette

SECONDARY SOURCES

Akenson, Donald H. *The Irish in Ontario: A Study in Rural History.* Kingston and Montreal: McGill-Queen's University Press, 1984.
– "Ontario: Whatever Happened to the Irish?" In *Canadian Papers in Rural History,* edited by Akenson, 3: 204–56. Gananoque, Ont.: Langdale Press, 1982.
Anderson, Richard. "'The Irrepressible Stampede': Tramps in Ontario, 1870–1880." *Ontario History* 84 (1992): 33–56.
Armstrong, F.H. "The Toronto Directories and the Negro Community in the Late 1840s." *Ontario History* 59 (1969): 111–19.
Arnold, Richard M. "Geography of Colonial Unrest in Upper Canada, 1820–1840." Master's thesis, York University, 1985.
Badè, William Frederic. *The Life and Letters of John Muir.* 2 vols. Boston: Houghton Mifflin, 1924.
Baker, Alan R.H., and Gideon Biger, eds. *Ideology and Landscape in Historical Perspective: Essays on the Meanings of Some Places in the Past.* Cambridge: Cambridge University Press, 1992.
Ball, Norman R. "The Technology of Settlement and Land Clearing in Upper Canada Prior to 1840." Ph.D. diss., University of Toronto, 1979.
– , ed. *Building Canada: A History of Public Works.* Toronto: University of Toronto Press, 1988.

Banks, Margaret. "Upper and Lower Canada or Canada West and East, 1841–1867." *Canadian Historical Review* 54 (1973): 473–80.

Barron, F.L. "The Genesis of Temperance in Ontario, 1828–1850." Ph.D. diss., University of Guelph, 1976.

Barron, Hal S. *Those Who Stayed Behind: Rural Society in Nineteenth-Century New England.* Cambridge: Cambridge University Press, 1984.

Baskerville, Peter. "Transportation, Social Change, and State Formation, Upper Canada, 1841–1864." In *Colonial Leviathan: State Formation in Mid-Nineteenth-Century Canada,* edited by Allan Greer and Ian Radforth, 230–56. Toronto: University of Toronto Press, 1992.

Becker, Carl L. *The Heavenly City of the Eighteenth-Century Philosophers.* New Haven: Yale University Press, 1932.

Bogue, Allan. *From Prairie to Corn Belt: Farming on the Illinois and Iowa Prairies in the Nineteenth Century.* Chicago: University of Chicago Press, 1963.

Bonnycastle, Richard H. *The Canadas in 1841.* 2 vols. London: Henry Colburn, 1842.

Bouchard, Gerard. "Family Reproduction in New Rural Areas: Outline of a North American Model." *Canadian Historical Review* 75 (1994): 475–510.

Brookes, Alan A., and Catharine A. Wilson. " 'Working Away' from the Farm: The Young Women of North Huron, 1910–1930." *Ontario History* 77 (1985): 281–300.

Brunger, Alan G. "The Distribution of English in Upper Canada 1851–1871." *Canadian Geographer* 30 (1986): 337–43.

– "The Distribution of Scots and Irish in Upper Canada, 1851–1871." *Canadian Geographer* 34 (1990): 250–8.

– "Geographical Propinquity among Pre-Famine Catholic Irish Settlers in Upper Canada." *Journal of Historical Geography* 8 (1982): 265–82.

Bylund, Erik. "Theoretical Considerations Regarding the Distribution of Settlement in North Sweden." *Geografiska Annaler* 42 (1960): 225–31.

Cameron, James M. "The Canada Company and Land Settlement as Resource Development in the Guelph Block." In *Perspectives on Landscape and Settlement in Nineteenth-Century Ontario,* edited by J. David Wood, 141–58. Carleton Library, no. 91. Toronto: McClelland & Stewart, 1975.

Canada. Act for Abolishing the Territorial Division of Upper Canada into Districts. 12 Victoria, c. 78.

– Act to Provide, by One General Law, for the Erection of Municipal Corporations. 12 Victoria, c. 81.

– Board of Registration and Statistics. *Census of the Canadas, 1851–2.* 2 vols. Quebec, 1853, 1855.

– *Census of Canada 1870–71.* Vol. 1. Ottawa, 1873.

– *Census of Canada 1870–71.* Vol. 4. *Censuses of Canada: 1665 to 1871.* Ottawa, 1876.

– House of Commons. *Debates.* 1887.

– *Journals of the Legislative Assembly of the Province of Canada.* 1841. Appendix T, "Population Returns of the Late Province of Upper Canada for the year 1840, etc."

– *Journals of the Legislative Assembly of the Province of Canada.* 1841. "Report of Select Committee to Which Were Referred the Petition of the Directors of the Toronto and Lake Huron Rail Road Company, and Other References." 16 Sept. 1841.

– *Journals of the Legislative Assembly of the Province of Canada.* 1842. Vol. 2. Appendix U, "Annual Report, for 1842, of the Commissioners of the Temporary Lunatic Asylum at Toronto."

– *Journals of the Legislative Assembly of the Province of Canada.* 1846. Appendix F, nos 16, 17, 19, "Report of Deputy Postmaster General (1840–41)" by T.A. Stayner, 1841.

– *Journals of the Legislative Assembly of the Province of Canada.* 1849. Appendix V.V.V., "Estimate of Certain Expenses of the Civil Government of the Province of Canada For the Year 1849, etc."

– *Journals of the Legislative Assembly of the Province of Canada.* 1849. Appendix Z.Z.Z., "Report of the Select Committee Appointed to Inquire Whether Any, and What Measures Can Be Adopted to Repress the Evils Growing Out of Intemperance."

– Patent Office. *Patents of Canada from 1824–49.* Vol. 1. Toronto, 1860.

– Patent Office. *Patents of Canada from 1849–55.* Vol. 2. Toronto, 1865.

Canada in the Years 1832, 1833, and 1834. By an Ex-Settler. Dublin: P.D. Hardy, 1835.

Careless, J. Maurice S. *Toronto to 1918: An Illustrated History.* Toronto: Lorimer & National Museums of Canada, 1984.

– *The Union of the Canadas: The Growth of Canadian Institutions, 1841–1857.* Toronto: McClelland & Stewart, 1967.

Clarke, John, H.W. Taylor, and W.R. Wightman. "Areal Patterns of Population Change in Southern Ontario 1831–1891: Core, Frontier and Intervening Space." *Ontario Geography* 12 (1978): 27–48.

Clarke, John, and John Buffone. "Social Regions in Mid-Nineteenth-Century Ontario." *Histoire sociale/Social History* 28 (1995): 193–217.

Clarke, John, and P.K. MacLeod. "Concentration of Scots in Rural Southern Ontario." *Canadian Cartographer* 11 (1974): 107–13.

Cook, Ramsay. *1492 and All That: Making a Garden Out of a Wilderness.* The Robarts Lecture, Oct. 1992. Toronto: The Robarts Centre for Canadian Studies, York University, 1993.

Courville, Serge. *Introduction à la géographie historique.* Ste-Foy: Les Presses de l'Université Laval, 1995.

– "Space, Territory, and Culture in New France: A Geographical Perspective." In *People, Places, Patterns, Processes: Geographical Perspectives on the Canadian Past,* edited by Graeme Wynn, 165–76. Toronto: Copp, Clark, Pitman, 1990.

– "Villages and Agriculture in the Seigneuries of Lower Canada: Conditions of a Comprehensive Study of Rural Quebec in the First Half of the Nineteenth Century." In *Canadian Papers in Rural History*, edited by Donald H. Akenson. 5: 123–9. Gananoque, Ont.: Langdale Press, 1986.

Cowgill, Donald O. "Transition Theory as General Population Theory." *Social Forces* 41 (1963): 270–4.

Cox, Thomas R., R.S. Maxwell, P.D. Thomas, and J.T. Malone. *This Well-Wooded Land: Americans and Their Forests from Colonial Times to the Present.* Lincoln: University of Nebraska Press, 1985.

Craig, Gerald H., ed. *Lord Durham's Report: An Abridgement of "Report on the Affairs of British North America."* Carleton Library, no. 1. Toronto: McClelland & Stewart, 1963.

Cronon, William. *Changes in the Land: Indians, Colonists, and the Ecology of New England.* New York: Hill & Wang, 1983.

Crosby, Alfred W. "The British Empire as a Product of Continental Drift." In *Germs, Seeds and Animals: Studies in Ecological History*, 62–81. Armonk, N.Y.: M.E. Sharpe, 1994.

– *Ecological Imperialism: The Biological Expansion of Europe, 900–1900.* Cambridge & New York: Cambridge University Press, 1986.

Cross, Michael S. "'The Dark Druidical Groves'; The Lumber Community and the Commercial Frontier in British North America to 1854." Ph.D. diss., University of Toronto, 1968.

– "1837: The Necessary Failure." In *Pre-Industrial Canada, 1760–1849*, edited by Michael S. Cross and Gregory S. Kealey. 141–58. Toronto: McClelland & Stewart, 1982.

Cruikshank, E.A., ed. *The Correspondence of Lieut. Governor John Graves Simcoe, with Allied Documents Relating to His Administration of the Government of Upper Canada.* 5 vols. Toronto: The Ontario Historical Society, 1923–31.

Curti, Merle. *The Making of an American Community: A Case Study of Democracy in a Frontier County.* Stanford: Stanford University Press, 1959.

Curtis, Bruce. *Building the Educational State: Canada West, 1836–1871.* London Ont.: Althouse Press, 1988.

Dickens, Charles. *American Notes for General Circulation.* 1842. Reprint, edited by John S. Whiteley and Arnold Goldman, Harmondsworth: Penguin Books, 1972.

Dickens, David R., and Andrea Fontana, eds. *Postmodernism and Social Inquiry.* New York & London: Guildford Press, 1994.

Eadie, James A. "The Napanee Mechanics' Institutes: The Nineteenth-Century Ontario Mechanics' Institute Movement in Microcosm." *Ontario History* 68 (1976): 209–21.

Easterlin, R.A. "Population Change and Farm Settlement in the Northern United States." *Journal of Economic History* 36 (1976): 45–83.

Economic Atlas of Ontario. Edited by William G. Dean, with cartography by

G.J. Matthews. Toronto: University of Toronto Press, 1969.

Ellis, Bernice Merrick. *Life on the Old Plank Road: Highway 88.* Bond Head, Ont.: Tecumseth & West Gwillimbury Historical Society, 1984.

Ennals, Peter. "Cobourg and Port Hope: The Struggle for Control of 'the Back Country'." In *Perspectives on Landscape and Settlement in Nineteenth-Century Ontario,* edited by J. David Wood, 182–95. Carleton Library, no. 91. Toronto: McClelland & Stewart, 1975.

Errington, Elizabeth Jane. *Wives and Mothers, School Mistresses and Scullery Maids: Working Women in Upper Canada, 1790–1840.* Montreal & Kingston: McGill-Queen's University Press, 1995.

Evernden, Neil. *The Social Construction of Nature.* Baltimore & London: Johns Hopkins University Press, 1992.

Fallis, Lawrence S., Jr. "The Idea of Progress in the Province of Canada: A Study in the History of Ideas." In *The Shield of Achilles: Aspects of Canada in the Victorian Age,* edited by W.L. Morton, 169–83. Toronto: University of Toronto Press, 1968.

Fingard, Judith. "The Poor in Winter: Seasonality and Society in Pre-Industrial Canada." In *Pre-Industrial Canada, 1760–1849,* edited by Michael S. Cross and Gregory S. Kealey, 62–78. Toronto: McClelland & Stewart, 1982.

Flader, Susan, ed. *The Great Lakes Forest.* Minneapolis: University of Minnesota Press, 1983.

Flaherty, David H., ed. *Essays in the History of Canadian Law.* Vol. 2. Toronto: University of Toronto Press, 1983.

Gad, G., and D. Holdsworth. "Streetscape and Society: The Changing Built Environment of King Street, Toronto." In *Patterns of the Past: Interpreting Ontario's History,* edited by Roger Hall, William Westfall, and Laura Sefton MacDowell, 174–205. Toronto: Dundurn Press, 1988.

Gagan, David. "Enumerator's Instructions for the Census of Canada 1852 and 1861." *Histoire sociale/Social History* 7 (1974): 355–65.

– *Hopeful Travellers: Families, Land, and Social Change in Mid-Victorian Peel County, Canada West.* Toronto: University of Toronto Press, 1981.

Garland, M.A., and J.J. Talman. "Pioneer Drinking Habits and the Rise of Temperance Agitation in Upper Canada Prior to 1840." In *Aspects of Nineteenth-Century Ontario: Essays Presented to James J. Talman,* edited by F.H. Armstrong, H.A. Stevenson, and J.D. Wilson, 171–93. Toronto: University of Toronto Press, 1974.

Gates, Lillian F. *Land Policies of Upper Canada.* Toronto: University of Toronto Press, 1968.

Gentilcore, R. Louis, and C.G. Head. *Ontario's History in Maps.* Toronto: University of Toronto Press, 1984.

Gidney, R.D. "Elementary Education in Upper Canada: A Reassessment." In *Education and Social Change: Themes from Ontario's Past,* edited by Michael B. Katz and Paul H. Mattingly, 3–26. New York: New York University Press, 1975.

Gilmour, James M. *Spatial Evolution of Manufacturing: Southern Ontario 1851–1891*. University of Toronto Department of Geography Research Publications, no. 10. Toronto: University of Toronto Press, 1972.

Glacken, Clarence. *Traces on the Rhodian Shore: Nature and Culture in Western Thought from Ancient Times to the End of the Eighteenth Century*. Berkeley: University of California Press, 1967.

Glazebrook, G.P. de T. *A History of Transportation in Canada*. Vol. 1, *Continental Strategy to 1867*. Carleton Library, no. 11. Toronto: McClelland & Stewart, 1964.

Godfrey, Charles M. *Medicine for Ontario. A History*. Belleville, Ont.: Mika Publishing, 1979.

Godley, John Robert. "A High Church View of Canada in the 1840s." In *Early Travellers in the Canadas 1791–1867*, edited by Gerald M. Craig, 142–50. Toronto: Macmillan, 1955.

Goheen, Peter G. "Canadian Communications Circa 1845." *Geographical Review* 77 (1987): 35–51.

– "The Changing Bias of Inter-Urban Communications in Nineteenth-Century Canada." *Journal of Historical Geography* 16 (1990): 177–96.

– *Victorian Toronto, 1850 to 1900: Pattern and Process of Growth*. Research Paper no. 127, Chicago: University of Chicago Department of Geography, 1970.

Goldie, John. *Diary of a Journey Through Upper Canada and Some of the New England States, 1819*. Privately published facsimile. [Toronto: William Tyrrell, 1967]

Gourlay, Robert F. *A Statistical Account of Upper Canada*. 1822. Reprint, abridged by S.R. Mealing. Carleton Library, no. 75. Toronto: McClelland & Stewart, 1974.

Graff, Harvey J. "Towards a Meaning of Literacy: Literacy and Social Structure in Hamilton, Ontario, 1861." In *Education and Social Change: Themes from Ontario's Past*, edited by Michael B. Katz and Paul H. Mattingly, 246–70. New York: New York University Press, 1975.

Griffiths, Tom, and Libby Robin, eds. *Ecology and Empire: Environmental History of Settler Societies*. Seattle: University of Washington Press, 1997.

Guelph Conference on the Conservation of Natural Resources of Ontario. *Conservation and Post-War Rehabilitation*. Toronto: Guelph Conference, 1942.

[Haldimand and Norfolk Counties.] *Illustrated Historical Atlas of the Counties of Haldimand [1877] and Norfolk [1879]*. Toronto: H.R. Page & Co., 1877 and 1879. Reprint (2 vols. in 1), Owen Sound, Ont.: Richardson, Bond & Wright, 1972.

Hardy, René. *La sidérurgie dans le monde rural. Les hauts fourneaux du Québec au XIXe siècle*. Québec: Les Presses de l'Université Laval, 1995.

Harris, R. Cole, and John Warkentin. *Canada Before Confederation: A Study in Historical Geography*. New York: Oxford University Press, 1974; Ottawa: Carleton University Press, 1991.

Hart, John Fraser. "The Spread of the Frontier and the Growth of Population."

In *Man and Cultural Heritage: Papers in Honor of Fred B. Kniffen,* edited by H.J. Walker and W.G. Haag, 73–81. Geoscience and Man, vol. 5. Baton Rouge: Louisiana State University Press, 1974.

Hay, Douglas. "Property, Authority, and the Criminal Law." In *Albion's Fatal Tree: Crime and Society in Eighteenth-Century England,* edited by Douglas Hay, Peter Linebaugh, John G. Rule, E.P. Thompson, and Cal Winslow, 17–63. New York: Pantheon Books, 1975.

Head, C. Grant. "An Introduction to Forest Exploitation in Nineteenth-Century Ontario." In *Perspectives on Landscape and Settlement in Nineteenth-Century Ontario,* edited by J. David Wood, 78–112. Carleton Library, no. 91. Toronto: McClelland & Stewart, 1975.

– "The Pine Lumber Industry of Ontario: The Lake Erie Frontier, 1845–1865." Paper presented at the Eastern Historical Geography Association meeting, Alexandria, Va., 1995.

Henripin, Jacques. *Trends and Factors of Fertility in Canada.* 1961 census monograph. Ottawa: Statistics Canada, 1972.

Hill, Daniel G. "Negroes in Toronto, 1793–1865." *Ontario History* 55 (1963): 73–91.

Historical Atlas of Canada. Edited by R. Cole Harris, with cartography by Geoffrey J. Matthews. Vol. 1, *From the Beginning to 1800.* Toronto: University of Toronto Press, 1987.

Historical Atlas of Canada. Edited by R. Louis Gentilcore, with cartography by Geoffrey J. Matthews. Vol. 2, *The Land Transformed, 1800–1891.* Toronto: University of Toronto Press, 1993.

Houston, Cecil J., and William J. Smyth. *Irish Emigration and Canadian Settlement: Patterns, Links, and Letters.* Toronto: University of Toronto Press, 1990.

– "The Orange Order and the Expansion of the Frontier in Ontario, 1830–1900." *Journal of Historical Geography* 4 (1978): 251–64.

Houston, Susan E., and Alison Prentice. *Schooling and Scholars in Nineteenth-Century Ontario.* Toronto: University of Toronto Press, 1988.

Howison, John. *Sketches of Upper Canada, Domestic, Local, and Characteristic.* Edinburgh: Oliver & Boyd, 1821.

Hudson, J. "Migration to an American Frontier." *Annals of the Association of American Geographers* 66 (1976): 242–65.

Inge, W.R. *The Idea of Progress.* The Romanes Lecture. Oxford: Clarendon Press, 1920.

Inwood, Kris, and Tim Sullivan. "Comparative Perspectives on Nineteenth-Century Growth: Ontario in the Great Lakes Region." In *Canadian Papers in Business History,* edited by Peter Baskerville, 71–101. Victoria, B.C.: The Public History Group, University of Victoria, 1993.

Jameson, Anna Brownell. *Winter Studies and Summer Rambles in Canada. Selections.* 1838. Reprint, New Canadian Library, no. 46. Toronto: McClelland & Stewart, 1965.

Jasen, Patricia. *Wild Things: Nature, Culture, and Tourism in Ontario, 1790–1914.* Toronto: University of Toronto Press, 1995.

Johnson, James K. *Becoming Prominent: Regional Leadership in Upper Canada, 1791–1841.* Kingston and Montreal: McGill-Queen's University Press, 1989.

Jones, Robert Leslie. *History of Agriculture in Ontario 1613–1880.* Toronto: University of Toronto Press, 1946.

Katz, Michael B. *The People of Hamilton, Canada West: Family and Class in a Mid-Nineteenth-Century City.* Cambridge: Harvard University Press, 1975.

Keefer, Thomas C. *The Philosophy of Railroads,* edited by H.V. Nelles. Toronto: University of Toronto Press, 1972.

Kelly, Kenneth. "The Agricultural Geography of Simcoe County, Ontario, 1820–1880." Ph.D. diss., University of Toronto, 1968.

– "Damaged and Efficient Landscapes in Rural Southern Ontario, 1880–1900." *Ontario History* 66 (1974): 1–14.

– "The Development of Farm Produce Marketing Agencies and Competition Between Marketing Centres in Eastern Simcoe County, 1850–1875." *Canadian Papers in Rural History,* edited by Donald H. Akenson. 1: 67–86. Gananoque, Ont.: Langdale Press, 1978.

– "The Transfer of British Ideas on Improved Farming to Ontario During the First Half of the Nineteenth Century." *Ontario History* 63 (1971): 103–11.

– "Wheat Farming in Simcoe County in the Mid-Nineteenth Century." *Canadian Geographer* 15 (1971): 95–112.

Kerr, D.G.G., ed. *A Historical Atlas of Canada.* Toronto: Thomas Nelson, 1960.

Kirk, D.W. "Southwestern Ontario: The Areal Pattern of Settlements in 1850." Ph.D. diss., Northwestern University, 1949.

Klotz, Otto. *A Review of the Special Report of the Minister of Education on the Mechanics' Institutes [in] Ontario.* Toronto: Willing & Williamson, 1881.

Knight, David B. *Choosing Canada's Capital: Conflict Resolution in a Parliamentary System.* Ottawa: Carleton University Press, 1991.

Lambert, Richard S., and Paul Pross. *Renewing Nature's Wealth: A Centennial History of the Public Management of Lands, Forests and Wildlife in Ontario 1763–1867.* Ontario Department of Lands and Forests, 1967.

Law, Howard. "'Self-Reliance is the True Road to Independence': Ideology and the Ex-Slaves in Buxton and Chatham." *Ontario History* 77 (1985): 107–21.

Lemon, James T. *Best Poor Man's Country: A Geographical Study of Early Southeastern Pennsylvania.* New York: Norton, 1976.

Leyburn, James G. *Frontier Folkways.* New Haven: Yale University Press, 1935.

Loucks, Orie L. "New Light on the Changing Forest." In *The Great Lakes Forest,* edited by Susan L. Flader, 17–32. Minneapolis: University of Minnesota Press, 1983.

Lower, Arthur R.M. *Great Britain's Woodyard: British America and the Timber Trade, 1763–1867.* Montreal and London: McGill-Queen's University Press, 1973.

– *The North American Assault on the Canadian Forest.* Toronto: Ryerson Press, 1938.

– *Settlement and the Forest Frontier in Eastern Canada.* Toronto: Macmillan, 1936.

Mackay, Robert W.S. *The Canada Directory ... to November, 1851.* Montreal: John Lovell, 1851.

Marr, W.L. "Tenant vs. Owner Occupied Farms in York County, Ontario, 1871." In *Canadian Papers in Rural History,* edited by Donald H. Akenson. 4: 50–71. Gananoque, Ont.: Langdale Press, 1984.

Marshall, John U. *The Location of Service Towns: An Approach to the Analysis of Central Place Systems.* University of Toronto Department of Geography Research Publications, no. 3. Toronto: University of Toronto Press, 1969.

– *The Structure of Urban Systems.* Toronto: University of Toronto Press, 1989.

– "Toward an Umland-based Model of City Sizes." In *Canada: Geographical Interpretations: Essays in Honour of John Warkentin,* edited by James R. Gibson, 231–54. North York, Ont.: Geography Department, Atkinson College, York University, 1993.

Marshall, John U., and W.R. Smith. "The Dynamics of Growth in a Regional Urban System: Southern Ontario, 1851–1971." *Canadian Geographer* 22 (1978): 22–40.

Martin, Robert Montgomery. *History of the Colonies of the British Empire.* 1843. Reprint, London: Dawsons of Pall Mall, 1967.

Matthews, W. Thomas. "Local Government and the Regulation of the Public Market in Upper Canada, 1800–1860: The Moral Economy of the Poor?" *Ontario History* 79 (1987): 297–326.

Mays, H.J. "'A Place to Stand': Families, Land and Permanence in Toronto Gore Township, 1820–1890." *Historical Papers 1980*: 185–211.

McCalla, Douglas. *Planting the Province: The Economic History of Upper Canada, 1784–1870.* Toronto: University of Toronto Press, 1993.

– "Railways and the Development of Canada West, 1850–1870." In *Colonial Leviathan: State Formation in Mid-Nineteenth-Century Canada,* edited by Allan Greer and Ian Radforth, 192–229. Toronto: University of Toronto Press, 1992.

McCuaig, J.D., and E.W. Manning. *Agricultural Land-Use Change in Canada: Process and Consequences.* Land Use series, no. 21. Ottawa: Environment Canada, 1982.

McIlwraith, Thomas F. "The Adequacy of Rural Roads in the Era Before Railways: An Illustration from Upper Canada." *Canadian Geographer* 14 (1970): 344–60.

– "British North America, 1763–1867." In *North America. The Historical Geography of a Changing Continent,* edited by Robert D. Mitchell and Paul A. Groves, 220–52. Totowa, N.J.: Rowman & Littlefield, 1987.

– "Freight Capacity and Utilization of the Erie and Great Lakes Canals before 1850." *Journal of Economic History* 36 (1976): 852–77.

– "The Logistical Geography of the Great Lakes Grain Trade, 1820–1850." Ph.D. diss., University of Wisconsin, 1973.

– *Looking for Old Ontario: Two Centuries of Landscape Change.* Toronto: University of Toronto Press, 1997.

McInnis, Marvin. "Childbearing and Land Availability: Some Evidence from Individual Household Data." In *Population Patterns in the Past*, edited by R.D. Lee, 201–27. New York: Academic Press, 1977.
– "Ontario Agriculture, 1851–1901: A Cartographic Overview." In *Canadian Papers in Rural History*, edited by Donald H. Akenson. 5: 290–301. Gananoque, Ont.: Langdale Press, 1986.
– "Perspectives on Ontario Agriculture, 1815–1930." In *Canadian Papers in Rural History*, edited by Donald H. Akenson. 8: 17–127. Gananoque, Ont.: Langdale Press, 1992.
McNeill, William H. *The Global Condition: Conquerors, Catastrophes, and Community*. Princeton: Princeton University Press, 1992.
McQuire, Scott. *Visions of Modernity: Representation, Memory, Time and Space in the Age of the Camera*. London: Sage, 1998.
Milani, Lois D. *Robert Gourlay, Gadfly: The Biography of Robert (Fleming) Gourlay, 1778–1863, Forerunner of the Rebellion in Upper Canada, 1837*. Thornhill, Ont.: Ampersand Press, 1971.
Mitchell, Robert D. *Commercialism and Frontier: Perspectives on the Early Shenandoah Valley*. Charlotteville: University Press of Virginia, 1977.
Moodie, Susanna. *Roughing it in the Bush; Or, Forest Life in Canada*. 1852. Reprint, Toronto: Coles, 1980.
Newell, Dianne. *Technology on the Frontier: Mining in Old Ontario*. Vancouver: University of British Columbia Press, 1986.
Nitkin, David. "Negro Colonization as a Response to Racism: A Historical Geography of the Southwestern Ontario Experience, 1830–1860." Master's thesis, York University, 1973.
Noel, Jan. *Canada Dry: Temperance Crusades before Confederation*. Toronto: University of Toronto Press, 1995.
Noel, S.J.R. *Patrons, Clients, Brokers: Ontario Society and Politics, 1791–1896*. Toronto: University of Toronto Press, 1990.
Norris, Darrell. "Household and Transiency in a Loyalist Township: The People of Adolphustown, 1784–1822." *Histoire sociale/Social History* 13 (1980): 399–415.
Ontario. Agricultural Commission. *Report of the Commissioners*. Vol. 2. Toronto, 1881.
– Department of Agriculture. *Report on the Reforestation of Waste Lands in Southern Ontario 1908*, by E.J. Zavitz. Toronto: King's Printer, 1909.
– Department of Planning and Development. *Humber Valley Watershed Report 1948*. Duplicated typescript edition. Toronto, 1948.
– Legislative Assembly. *Appendix to the Report of the Ontario Bureau of Industries 1896*. Vol. 29, pt 9 in *Sessional Papers of the Legislative Assembly, 1897*. Toronto: Queen's Printer, 1897.
– "Report of the Fruit Growers' Association of Ontario for 1869." Appendix C in *Report of the Commissioner of Agriculture and Arts for the Province of Ontario For the Year 1869*. Toronto, 1870.

Ontario Association of Agricultural Societies. *The Story of Ontario Agricultural Fairs and Exhibitions, 1792–1967*. Picton, 1967.

Osborne, Brian S. "Frontier Settlement in Eastern Ontario in the Nineteenth Century: A Study in Changing Perceptions of Land and Opportunity." In *The Frontier: Comparative Studies*, edited by D.H. Miller and J.O. Steffen, 201–25. Norman: University of Oklahoma Press, 1977.

– "Trading on a Frontier: The Function of Peddlers, Markets and Fairs in Nineteenth-Century Ontario." In *Canadian Papers in Rural History*, edited by Donald H. Akenson. 2: 59–81. Gananoque, Ont.: Langdale Press, 1980.

– "Wills and Inventories: Records of Life and Death in a Developing Society." *Families* 19 (1980): 235–47.

Osborne, Brian S., and Donald Swainson. *Kingston: Building on the Past*. Westport, Ont.: Butternut Press, 1988.

Owram, Doug. *Promise of Eden: The Canadian Expansionist Movement and the Idea of the West, 1856 to 1900*. Toronto: University of Toronto Press, 1980.

Park, Deborah Carter, and J. David Wood. "Poor Relief and the County House of Refuge System in Ontario, 1880–1911." *Journal of Historical Geography* 18 (1992): 439–55.

Pattison, Irma E. *Historical Chronology of Highway Legislation in Ontario, 1774–1961*. [Toronto]: Ontario Department of Highways, 1964.

Prentice, Alison. *The School Promoters: Education and Social Class in Mid-Nineteenth-Century Upper Canada*. Toronto: McClelland & Stewart, 1977.

Primack, Martin L. "Land Clearing Under Nineteenth Century Techniques." *Journal of Economic History* 22 (1962): 484–97.

[Proudfoot, William.] "The Proudfoot Papers." *Ontario Historical Society Papers and Records* 29 (1933): 141–59.

Radcliff, Thomas, ed. *Authentic Letters From Upper Canada, Including an Account of Field Sports by Thomas William Magrath*. Toronto: Macmillan, 1952.

Read, Colin, and R.J. Stagg, eds. *The Rebellion of 1837 in Upper Canada: A Collection of Documents*. Toronto: The Champlain Society and The Ontario Heritage Foundation, 1985.

Report of the London and Port Stanley Railway Company. 31 July 1857. Canadian Institute for Historical Microreproductions, microfiche no. 37189.

La Rochefoucauld-Liancourt's Travels in Canada 1795. Edited by W.R. Riddell. *Thirteenth Report of the Bureau of Archives of the Province of Ontario, 1916*. Toronto: King's Printer, 1917.

Rogers, Ian MacF. *The Law of Municipal Corporations*. 2nd ed. Toronto: Carswell Co., 1971.

Rorke, Richard. *Forty Years in the Forest: Reminiscences from the Pen of a Backwoodsman, 1820–1868*. Edited by P.E.K. Armstrong. Tecumseth & West Gwillimbury Historical Society, 1987.

Russell, Peter. "Forest into Farmland: Upper Canada Clearing Rates, 1822–1839." *Agricultural History* 57 (1983): 326–39.

Sauer, Carl O. "The Agency of Man on the Earth." In *Man's Role in Changing the Face of the Earth*, edited by William L. Thomas, 49–69. Chicago: University of Chicago Press, 1956.

Saunders, Robert E. "What Was the Family Compact?" In *Historical Essays on Upper Canada*, edited by J.K. Johnson, 122–39. Carleton Library, no. 82. Toronto: McClelland & Stewart, 1975.

Schama, Simon. *Landscape and Memory*. Toronto: Random House, Vintage, 1996.

[Simcoe County.] *Illustrated Historical Atlas of the County of Simcoe*. Toronto: H. Belden & Co., 1881. Reprint, Port Elgin, Ont.: Ross Cumming, 1970.

Smaller, Harry. "Teachers and Schools in Early Ontario." *Ontario History* 85 (1993): 291–307.

Smith, D.A. *At the Forks of the Grand*. Paris: Walker Press, 1952.

Smith, W. Randy. *Aspects of Growth in a Regional Urban System: Southern Ontario 1851–1921*. Geographical Monographs, no. 12. Toronto: Atkinson College, York University, 1982.

Smith, William H. *Canada: Past, Present and Future. Being a Historical, Geographical, Geological and Statistical Account of Canada West*. 2 vols. Toronto: Maclear, 1852.

– *The Canadian Gazetteer Comprising Statistical and General Information*. 1846. Reprint, Toronto: Coles, 1972.

Snell, Elizabeth A. *Wetland Distribution and Conversion in Southern Ontario*. Working Paper no. 48. [Ottawa]: Inland Waters and Lands Directorate, Environment Canada, 1987.

Sokoloff, Kenneth L. "Invention, Innovation, and Manufacturing Productivity Growth in the Antebellum Northeast." In *American Economic Growth and Standards of Living before the Civil War*, edited by Robert E. Gallman and J.J. Wallis, 345–84. Chicago: University of Chicago Press, 1992.

Spelt, Jacob. *Urban Development in South-Central Ontario*. Carleton Library, no. 57. Toronto: McClelland & Stewart, 1972.

Splane, Richard B. *Social Welfare in Ontario, 1791–1893: A Study of Public Welfare Administration*. Toronto: University of Toronto Press, 1965.

Spragge, George W. "The Districts of Upper Canada, 1788–1849." *In Profiles of a Province: Studies in the History of Ontario*, 34–42. Toronto: Ontario Historical Society, 1967.

St-Denis, Guy. "An Erie Canal for Western Upper Canada: A Forgotten Episode in Ontario's Transportation Evolution." *Ontario History* 85 (1993): 231–50.

Talman, J.J. "Agricultural Societies of Upper Canada." *Ontario Historical Society Papers and Records* 27 (1931): 545–52.

– "Travel in Ontario Before the Coming of the Railway." *Ontario Historical Society Papers and Records* 29 (1933): 85–102.

Thompson, R.L. *Wiring a Continent: The History of the Telegraph Industry in the United States 1832–1866*. Princeton: Princeton University Press, 1947.

Thomson, William. "A Tradesman's Travels." In *Early Travellers in the Canadas, 1791–1867*, edited by Gerald Craig, 132–41. Toronto: Macmillan, 1955.

Traill, Catharine Parr. *The Backwoods of Canada.* 1836. Reprint, New Canadian Library, no. 51. Toronto: McClelland & Stewart, 1966.

Trout, J.M., and E. Trout. *The Railways of Canada for 1870–1.* Toronto: The Monetary Times, 1871.

U.S. Congress. Senate. Secretary of the Treasury. *Report on the Trade, Commerce, and Resources of the British North American Colonies,* by J.D. Andrews. 31st Congress, 2nd sess., 1851. Ex. Doc. 23.

Vernon, F. "The Development of Adult Education in Ontario, 1790–1900." D.Ed. diss., University of Toronto, 1969.

Wallerstein, Immanuel. *The Modern World-System.* Vol. 3, *The Second Era of Great Expansion of the Capitalist World-Economy, 1730–1840s.* San Diego: Academic Press, 1989.

Walsh, Lorena S. "Consumer Behavior, Diet, and the Standard of Living in Late Colonial and Early Antebellum America, 1770–1840." In *American Economic Growth and Standards of Living Before the Civil War,* edited by Robert E. Gallman and J.J. Wallis, 217–64. Chicago: University of Chicago Press, 1992.

Warkentin, John. *Canada: A Regional Geography.* Scarborough: Prentice-Hall, 1997.

Weaver, W.F. "Ontario Surveys and the Land Surveyor." *Canadian Geographical Journal* 32 (April 1946): 180–91.

Whebell, C.F.J. "Corridors: A Theory of Urban Growth." *Annals of the Association of American Geographers* 59 (1969): 1–26.

– "Robert Baldwin and Decentralization 1841–9." In *Aspects of Nineteenth-Century Ontario,* edited by F.H. Armstrong, H.A. Stevenson, and J.D. Wilson, 48–64. Toronto: University of Toronto Press, 1974.

– "The Upper Canada District Councils Act of 1841 and British Colonial Policy." *Journal of Imperial and Commonwealth History* 17 (1989): 185–209.

Wicksteed, G.W. *Table of Provincial Statutes in Force or Which have been in Force in Upper Canada, in their Chronological Order ... to the End of the Session of 1856.* Toronto: Queen's Printer, 1856.

Widdis, Randy William. "Speculation and the Surveyor: An Analysis of the Role Played by Surveyors in the Settlement of Upper Canada." *Histoire sociale/Social History* 15 (1982): 443–58.

Wilkinson, Anne. *Lions in the Way: A Discursive History of the Oslers.* Toronto: Macmillan, 1956.

Williams, Michael. *Americans and Their Forests: A Historical Geography.* Cambridge: Cambridge University Press, 1989.

Wilson, Alan. *The Clergy Reserves of Upper Canada: A Canadian Mortmain.* Toronto: University of Toronto Press, 1968.

Winder, Gordon M. "Before the Corporation and Mass Production: The Licensing Regime in the Manufacturing of North American Harvesting Machinery, 1830–1910." *Annals of Association of American Geographers* 85 (1995): 521–52.

Winearls, Joan. *Mapping Upper Canada 1780–1867: An Annotated Bibliography.* Toronto: University of Toronto Press, 1992.

Wood, J. David. "The Arks of Absalom Shade." *Inland Seas* 15 (1959): 180–5.

– "Moraine and Metropolis: The Oak Ridges Moraine and the Greater Toronto Area." *International Journal of Environmental Studies* 39 (1991): 45–53.

– "Population Change on an Agricultural Frontier: Upper Canada, 1796 to 1841." In *Patterns of the Past: Interpreting Ontario's History,* edited by Roger Hall, William Westfall, and Laura Sefton MacDowell, 55–77. Toronto: Dundurn Press, 1988.

– "The Population of Ontario: A Study of the Foundation of a Social Geography." In *A Social Geography of Canada,* edited by Guy M. Robinson, 92–137. Toronto: Dundurn Press, 1991.

– "Reflections on Administering a Forested Agricultural Frontier." *Canadian Geographer* 36 (1992): 254–60.

– "The Settlers and the Land: Pioneer Experience in the Home District." *Families* 14 (1975): 108–25.

– "The Stage is Set: Dumfries Township, 1816." *Waterloo Historical Society Annual Volume* 48 (1961): 40–50.

– "The Woodland-Oak Plains Transition Zone in the Settlement of Western Upper Canada." *Canadian Geographer* 5 (1961): 43–7.

Wycoff, William. *The Developer's Frontier: The Making of the Western New York Landscape.* New Haven: Yale University Press, 1988.

Wynn, Graeme. "Notes on Society and Environment in Old Ontario." *Journal of Social History* 13 (1979): 49–65.

[York and Simcoe Counties.] *Illustrated Historical Atlas of the County of York and the Township of West Gwillimbury and Town of Bradford in the County of Simcoe.* Toronto: Miles & Co., 1878. Reprint, Mika Silk Screening Ltd., 1972.

Zelinsky, Wilbur. *The Cultural Geography of the United States.* Rev. ed. Englewood Cliffs: Prentice-Hall, 1992.

Zeller, Suzanne. *Inventing Canada: Early Victorian Science and the Idea of a Transcontinental Nation.* Toronto: University of Toronto Press, 1987.

Index